USA009026
Basic

NASA
SPACE SHUTTLE CREW ESCAPE SYSTEMS HANDBOOK

Space Flight Operations Contract

Crew Escape Systems 21002

January 17, 2005

Contract NAS9-20000

United Space Alliance

©2012 Periscope Film LLC
All Rights Reserved
ISBN#978-1-937684-78-5
www.PeriscopeFilm.com

USA009026
Basic

Crew Escape Systems 21002

Prepared by

___*Original signature obtained*___
J. Lynn Coldiron, Book Manager
USA/Crew Escape

Approved by

___*Original signature obtained*___
Adam Flagan, Escape Technical Lead
USA/Crew Systems Group Lead

___*Original signature obtained*___
M. Jude Alexander, Lead
USA/Photo/TV/Crew Systems Group

Contract NAS9-20000

USA009026
Basic

REVISION LOG

Rev. letter	Change no.	Description	Date
Basic		Supersedes SFOC-FL0236	01/17/2005

USA009026
Basic

LIST OF EFFECTIVE PAGES

The status of all pages in this document is shown below:

Page No.	Change No.
i – vi	Basic
1-1 – 1-2	Basic
2-1 – 2-41	Basic
3-1 – 3-41	Basic
4-1 – 4-5	Basic
5-1 – 5-45	Basic
A-1 – A-3	Basic
B-1 – B-3	Basic

USA009026
Basic

PREFACE

This document was prepared by the United Space Alliance under contract to the Mechanical, Booster, and Maintenance Systems Branch, Systems Division, NASA, Lyndon B. Johnson Space Center, Houston, Texas. Documentation support was provided by Hernandez Engineering, Inc., Space Flight Operations Contract (HEI-SFOC).

Information contained in this document is provided for the use and training of crewmembers and for use by escape instructors and others who need to know about equipment, systems, and procedures relating to orbiter emergency egress and crew rescue.

The document describes and explains the use of crew-worn equipment and orbiter hardware. It also identifies the escape modes and the orbiter crew response in each mode. It is intended to be a workbook to which notes and additional information presented in classroom sessions can be added.

Mission-specific data will sometimes differ from the material contained in this document. In that case, flight supplements, Crew Compartment Configuration Drawings (CCCDs), and escape instructors should be consulted.

Additions and revisions to the document will be issued as required. Periodic changes will be distributed per the distribution list. Those not on distribution, may call for updates. For copies of the document, call the Mission Operations Library at (281) 244-7060, or pick up copies in person at the library, B-4S, Room 1303.

Comments, questions, or changes to this document should be directed to DX33, Crew Systems Section, (281) 483-0651 or 244-0092.

CONTENTS

Section		Page
1.0	**INTRODUCTION**	**1-1**
1.1	OVERVIEW	1-1
1.1.1	Introduction	1-1
1.1.2	Purpose	1-1
1.1.3	Objectives	1-2
2.0	**CREW-WORN EQUIPMENT**	**2-1**
2.1	OVERVIEW	2-1
2.1.1	Introduction	2-1
2.1.2	Purpose	2-1
2.1.3	Objectives	2-1
2.2	ADVANCED CREW ESCAPE SUIT (ACES)	2-2
2.2.1	Introduction	2-2
2.2.2	ACES Features	2-5
2.2.3	ACES O_2 Flow, Pressure System	2-5
2.2.4	Pressure System Components	2-7
2.2.5	Anti-g Protection	2-9
2.2.6	O_2 Manifold/G-Suit Controller Assembly	2-10
2.2.7	ACES Gloves	2-12
2.2.8	Boots	2-13
2.2.9	Crewmember Identification Patches	2-13
2.3	HELMET, COMM CARRIER ASSEMBLY, HEADSET INTERFACE UNIT	2-14
2.3.1	Introduction	2-14
2.3.2	Helmet Features	2-14
2.3.3	CCA	2-17
2.3.4	CCA Components	2-17
2.3.5	HIU	2-18
2.4	COOLING SYSTEMS	2-19
2.4.1	Introduction	2-19
2.4.2	TELCU/ICU Cooling	2-19
2.4.3	LCG	2-21
2.5	PARACHUTE HARNESS, PERSONAL PARACHUTE ASSEMBLY	2-22
2.5.1	Introduction	2-22
2.5.2	Parachute Harness	2-22
2.5.3	Parachute Assembly	2-26
2.5.4	Parachute Deployment	2-29
2.6	RESCUE/SURVIVAL GEAR	2-31
2.6.1	Introduction	2-31
2.6.2	Gear in Left Leg Suit Pocket	2-32
2.6.3	Gear in Right Leg Suit Pocket	2-34

Section			Page
	2.6.4	Shroudline Cutter/Knife	2-36
	2.6.5	Gear, Parachute Harness	2-36
	2.6.6	Gear in PPA/Life Raft	2-38
3.0	**ORBITER HARDWARE**		**3-1**
	3.1	OVERVIEW	3-1
	3.1.1	Introduction	3-1
	3.1.2	Purpose	3-1
	3.1.3	Objectives	3-1
	3.2	CREW SEATS	3-2
	3.2.1	Description	3-2
	3.2.2	Seat Functions and Features	3-6
	3.3	EMERGENCY EGRESS NET (TRAMPOLINE)	3-7
	3.3.1	Description	3-7
	3.3.2	Trampoline Attachment	3-7
	3.3.3	Closeout Nets	3-10
	3.3.4	Trampoline Stowage	3-10
	3.4	SIDE HATCH	3-12
	3.4.1	Introduction	3-12
	3.4.2	Physical Features	3-12
	3.4.3	Functional Features	3-13
	3.4.4	Normal Side Hatch Opening	3-13
	3.4.5	Depress, Hatch Jettison	3-14
	3.4.6	Side Hatch Opening by Rescue Personnel	3-16
	3.5	WINDOW 8 ESCAPE PANEL	3-18
	3.5.1	Introduction	3-18
	3.5.2	Panel Pyrotechnics	3-19
	3.5.3	Escape Panel Jettison	3-21
	3.5.4	Manual Opening with Prybar	3-22
	3.5.5	Opening by Rescue Personnel	3-23
	3.5.6	Cut-in Area	3-26
	3.6	DESCENT CONTROL DEVICE (SKY GENIE)	3-26
	3.6.1	Introduction	3-26
	3.6.2	Functional Features	3-27
	3.6.3	Sky Genie Use	3-27
	3.6.4	Sky Genie Use, Side Hatch	3-28
	3.6.5	Sky Genie Use, Window 8 Escape Panel	3-29
	3.7	ESCAPE POLE	3-30
	3.7.1	Introduction	3-30
	3.7.2	Functional Features	3-30
	3.7.3	Normal Escape Pole Deployment	3-31
	3.7.4	Manual Escape Pole Deployment	3-32
	3.7.5	Magazine and Lanyards	3-32
	3.7.6	Lanyard Use	3-33

Section | **Page**

	3.8	EMERGENCY EGRESS SLIDE SYSTEM	3-33
	3.8.1	Introduction	3-33
	3.8.2	Physical Characteristics	3-34
	3.8.3	Slide Deployment	3-35
	3.8.4	Slide Deploy, Hatch On	3-36
	3.8.5	Slide Deploy, Hatch Off	3-40
4.0	**NOMINAL CREW SEAT PROCEDURES**		**4-1**
	4.1	OVERVIEW	4-1
	4.1.1	Introduction	4-1
	4.1.2	Purpose	4-1
	4.1.3	Objectives	4-1
	4.2	NOMINAL SEAT INGRESS/EGRESS	4-1
	4.2.1	Prelaunch Seat Ingress	4-2
	4.2.2	Postinsertion Seat Egress	4-4
	4.2.3	Deorbit Prep Seat Ingress	4-4
	4.2.4	Nominal Seat/Orbiter Egress	4-5
5.0	**EMERGENCY EGRESS AND ESCAPE**		**5-1**
	5.1	OVERVIEW	5-1
	5.1.1	Introduction	5-1
	5.1.2	Purpose	5-1
	5.1.3	Objectives	5-1
	5.2	EMERGENCY EGRESS/ESCAPE MODES	5-2
	5.2.1	Introduction	5-2
	5.2.2	Assumptions	5-3
	5.2.3	Emergency Egress Cue Card	5-4
	5.3	MODE I – UNASSISTED PAD EGRESS/ESCAPE	5-5
	5.3.1	Introduction	5-5
	5.3.2	Slidewire Basket System	5-5
	5.3.3	Sequence of Action	5-7
	5.3.4	Armored Personnel Carrier	5-11
	5.3.5	Mode I Egress Cue Card Procedures	5-12
	5.4	MODES II, III, IV – ASSISTED PAD EGRESS/ESCAPE	5-15
	5.4.1	Introduction	5-15
	5.4.2	Egress Conditions	5-15
	5.4.3	Sequence of Action	5-15
	5.4.4	SSME Shutdown	5-17
	5.5	MODE V – UNASSISTED POSTLANDING EGRESS/ESCAPE	5-18
	5.5.1	Introduction	5-18
	5.5.2	Egress Routes, Means	5-19
	5.5.3	Hatch-on Mode V Slide	5-20
	5.5.4	Hatch-Jettison Mode V Slide Egress	5-22

Section			Page
	5.5.5	Window 8 Escape Panel Egress with Sky Genie	5-25
	5.5.6	Mode V Cue Card Procedures	5-27
	5.5.7	Mode V Convoy Crew Positioning	5-30
	5.5.8	Postlanding Loss of Orbiter Comm	5-30
5.6		MODES VI, VII – ASSISTED POSTLANDING EGRESS/ESCAPE	5-31
	5.6.1	Introduction	5-31
	5.6.2	Mode VI Convoy Positioning	5-32
5.7		MODE VIII – BAILOUT	5-32
	5.7.1	Introduction	5-32
	5.7.2	Bailout Sequence of Action	5-33
	5.7.3	Mode VIII Bailout Cue Card Procedures	5-35
	5.7.4	Parachute Deploy Sequence	5-39
	5.7.5	Sequence of Action When Landing in Water	5-41
	5.7.6	Contingency Landing Site Resources	5-43
	5.7.7	Launch Contingency Bailout Area Rescue	5-43

Appendix

A	ACRONYMS AND ABBREVIATIONS	A-1
B	CREW ESCAPE LESSONS	B-1

TABLES

Table		Page
2-1	Crewmember Color and Letter	2-13
2-2	PPA Components	2-28
5-1	Escape Modes	5-2
5-2	Mode I Crew Egress	5-7
5-3	Prelaunch	5-13
5-4	Egress Route and Means Used	5-19
5-5	Postlanding	5-29
5-6	Light Signals	5-31
5-7	Description	5-32
5-8	Mode VIII Crew Egress	5-33
5-9	Bailout	5-37
5-10	Deployment Sequence	5-39
5-11	Water Landing	5-41
5-12	Contingency Landing Sites and Resources	5-43
5-13	KSC Area SAR Response	5-44
5-14	ECAL/TAL Area SAR Response	5-44

FIGURES

Figure		
2-1	ACES, front and back views	2-4
2-2	ACES oxygen flow and pressure system	2-6
2-3	Dual suit controller	2-7
2-4	G-Suit	2-10
2-5	O_2 manifold/g-suit controller	2-10
2-6	ACES gloves	2-12
2-7	Helmet	2-14
2-8	Neck ring	2-15
2-9	The CCA	2-17
2-10	HIU	2-18
2-11	TELCU/ICU cooling system	2-20
2-12	LCG	2-22
2-13	Parachute harness (upper portion)	2-23
2-14	Harness	2-24
2-15	Frost fitting and riser	2-25
2-16	Ejector snaps	2-26
2-17	Personal parachute assembly	2-27
2-18	Automatic parachute deploy sequence	2-29
2-19	Manual parachute deploy sequence	2-30
2-20	Automatic parachute deploy sequence using the red knob	2-31
2-21	Rescue/survival gear, left leg suit pocket (packet A)	2-32

Figure		Page
2-22	Exposure mittens	2-34
2-23	Rescue/survival gear, right leg suit pocket (packet B)	2-34
2-24	EOS and O_2 Regulator Valve	2-37
2-25	Rescue/survival gear, PPA/life raft	2-39
2-26	Bailing pump	2-41
3-1	CDR and PLT seats	3-3
3-2	Mission specialist and payload specialist seats	3-3
3-3	5-point Restraint	3-4
3-4	Horizontal and vertical adjustment	3-4
3-5	Quick disconnect fittings	3-5
3-6	Emergency egress step	3-5
3-7	Recumbent seat kit frames	3-6
3-8	MS (PS) seats installed on RSK	3-6
3-9	Emergency egress net (Trampoline)	3-8
3-10	Trampoline attachment points – ceiling	3-8
3-11	Trampoline attachment points – deck	3-9
3-12	Ratchet assembly	3-9
3-13	Nomex extension and snap hooks	3-10
3-14	Trampoline – stowed	3-11
3-15	Orbiter side hatch, interior view	3-12
3-16	External side hatch opening/closing device	3-17
3-17	Window 8 escape panel	3-19
3-18	Escape panel jettison handle, panel C2	3-20
3-19	Escape panel pry locations	3-22
3-20	Prybar positioning	3-23
3-21	External T-handle initiator	3-24
3-22	Cut-in area	3-26
3-23	Sky genie components	3-27
3-24	Sky genie use	3-28
3-25	Deployed escape pole (looking aft)	3-30
3-26	Escape pole deployment	3-31
3-27	Magazine and lanyards	3-32
3-28	Emergency egress slide deployed (side hatch opened normally)	3-34
5-1	Emergency egress cue card	5-4
5-2	Overall view of slidewire basket system	5-6
5-3	Primary and alternate routes to baskets	5-9
5-4	M-113 armored personnel carrier	5-12
5-5	Mode I pad egress cue card procedures	5-12
5-6	Mode V postlanding egress cue card procedures	5-28
5-7	Convoy positioning, Mode V (and Mode VI)	5-30
5-8	Mode VIII bailout cue card procedures	5-36
5-9	SAR Recovery Posture	5-45
5-10	KSC area SAR recovery posture	5-45

USA009026
Basic

1.0 INTRODUCTION

1.1 OVERVIEW

1.1.1 Introduction

The crew escape systems facilitate safe and expeditious crew egress and escape from the orbiter in an emergency during the following mission phases:

a. Prelaunch

b. In flight (bailout)

c. Postlanding

The crew escape systems include:

a. Equipment worn by the crew; for example:

   1. Advanced Crew Escape Suit (ACES)

   2. Parachute harness and parachute assembly

   3. Rescue and survival equipment

b. Orbiter hardware for example:

   1. Side hatch jettisoning system

   2. Window 8 escape panel

   3. Descent control devices (sky genies)

   4. Escape pole (for bailout)

   5. Egress slide

1.1.2 Purpose

This document describes the crew escape systems, the operation of these systems, and their use during specific emergency or contingency situations. It also discusses systems associated with crew emergency egress, including orbiter nominal seating, the Recumbent Seat Kit (RSK), seat ingress/egress, and the side hatch.

1.1.3 Objectives

On completing this document, you will be able to

a. Identify the components of:

    1. Crew-worn escape equipment

    2. Orbiter and launch pad escape hardware

    3. Associated equipment and hardware

b. Identify the functional features of this equipment and hardware.

c. Identify the various emergency escape modes and crew action in each mode.

d. Know the correct sequence in bailout.

e. Identify Search and Rescue (SAR) resources for crew bailout.

In This Document

Chapters describe the crew escape systems and their use as follows:

Chapter	Page
Crew-Worn Equipment	2-1
Orbiter Hardware	3-1
Nominal Crew Seat Procedures	4-1
Emergency Egress and Escape	5-1

2.0 CREW-WORN EQUIPMENT

2.1 OVERVIEW

2.1.1 Introduction

The orbiter crewmembers wear equipment and gear that facilitate quick and safe egress/escape in an emergency occurring prelaunch, in flight, or postlanding. The crew-worn equipment and gear include the pressure suit, helmet, parachute, harness, rescue aids, and survival aids.

2.1.2 Purpose

This chapter describes the components, functional features, and operation and use of crew-worn emergency egress/escape equipment and associated hardware.

2.1.3 Objectives

On completing this chapter, you will be able to identify:

a. The components and functional features of:

1. Advanced Crew Escape Suit (ACES)
2. Helmet, Comm Carrier Assembly, Headset Interface Unit
3. Suit cooling systems
4. Parachute harness and parachute assembly
5. Rescue and survival equipment

b. The operation and use of this equipment and gear

In This Chapter

Sections describe the crew-worn escape gear as follows:

Section	Page
Advanced Crew Escape Suit (ACES)	2-2
Helmet, Comm Carrier Assembly, Headset Interface Unit	2-14
Cooling Systems	2-19
Parachute Harness, Personal Parachute Assembly	2-22
Rescue/Survival Gear	2-31

2.2 ADVANCED CREW ESCAPE SUIT (ACES)

2.2.1 Introduction

All orbiter crewmembers wear a protective suit during launch and entry. The crewmembers don and doff their suits as follows:

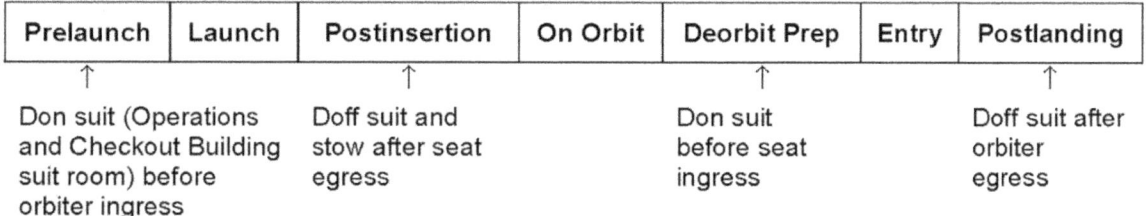

Prelaunch	Launch	Postinsertion	On Orbit	Deorbit Prep	Entry	Postlanding
↑		↑		↑		↑
Don suit (Operations and Checkout Building suit room) before orbiter ingress		Doff suit and stow after seat egress		Don suit before seat ingress		Doff suit after orbiter egress

Suit functions

a. The suit is designed to protect crewmembers from the following:

1. Loss of cabin pressure and/or orbiter oxygen at altitudes to 100,000 ft.

2. Pooling of blood in the lower body after prolonged exposure to microgravity.

3. Cold atmospheric and/or water temperatures after bailing out.

4. Fire

5. Contaminated atmosphere during flight or postlanding egress.

b. The suit delivers O_2 at the proper pressure and in the proper quantity for breathing and pressurization.

Note: A "hard pressure" is experienced when cabin pressure equals an altitude of ~55,000 ft. (In an in-flight emergency, this would be a good indication to the crew that bailout is not yet safe.)

At ~35,000 ft, the suit is unpressurized.

Suit Features

a. Full pressure vessel providing an "atmosphere of protection"

b. Protection up to 100,000 ft indefinitely

c. Full pressurization at 3.5 psi (relief valve operates at 5 psi)

d. A single bladder made of a single layer of nylon fabric laminated to Gore-Tex (which stays cooler)

e. Single neck dam dividing the helmet volume from the suit volume

f. Six pressure seals:

    1. Helmet visor

    2. Neck dam

    3. Main suit zipper

    4. Bioinstrumentation Passthrough (BIP) plug

    5. Right glove

    6. Left glove

g. Nonintegrated g-suit

h. Gloves required for suit pressurization

    1. Gloves connect to ACES via wrist disconnects

    2. Gloves must be worn to keep water out

i. Loose fit; adequate mobility.

The Advanced Crew Escape Suit (ACES) is shown in Figure 2-1 below.

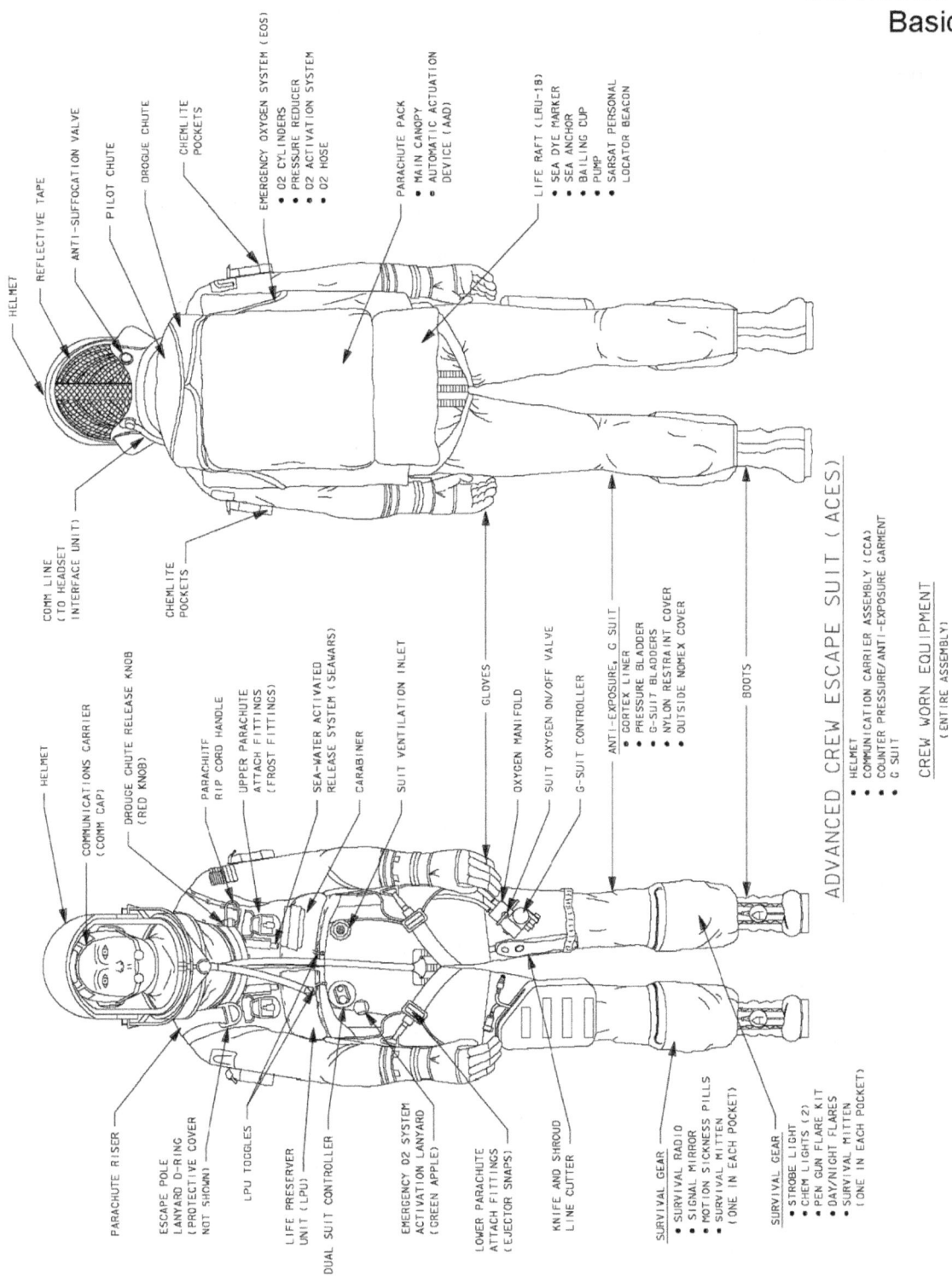

Figure 2-1. ACES, front and back views

2.2.2 ACES Features

The ACES has the following features:

a. A full pressure suit, the air pressure exerts direct pressure on the body. The orange outer garment of flame-retardant Nomex covers the single pressure bladder.

b. The detachable anti-g suit can be demated before launch and remated before entry.

c. The gloves are attached to the suit by disconnects and pressurize when the suit is pressurized.

d. The waterproof main zipper seals out water.

e. The vent connection has a flapper valve to prevent water from entering.

f. The single neck dam separates the helmet volume from the suit volume. The neck dam also prevents water from entering the suit when the helmet visor is open.

g. The neck ring tiedown is used to adjust the position of the neck ring.

Subsequent subsections describe the following ACES features in more detail:

a. ACES O_2 flow, Pressure system

b. Pressure system components

c. Anti-g protection

d. O_2 manifold/g-suit controller assembly

e. Cooling

f. Gloves, boots

2.2.3 ACES O_2 Flow, Pressure System

The ACES O_2 flow and pressure system is shown in Figure 2-2 below.

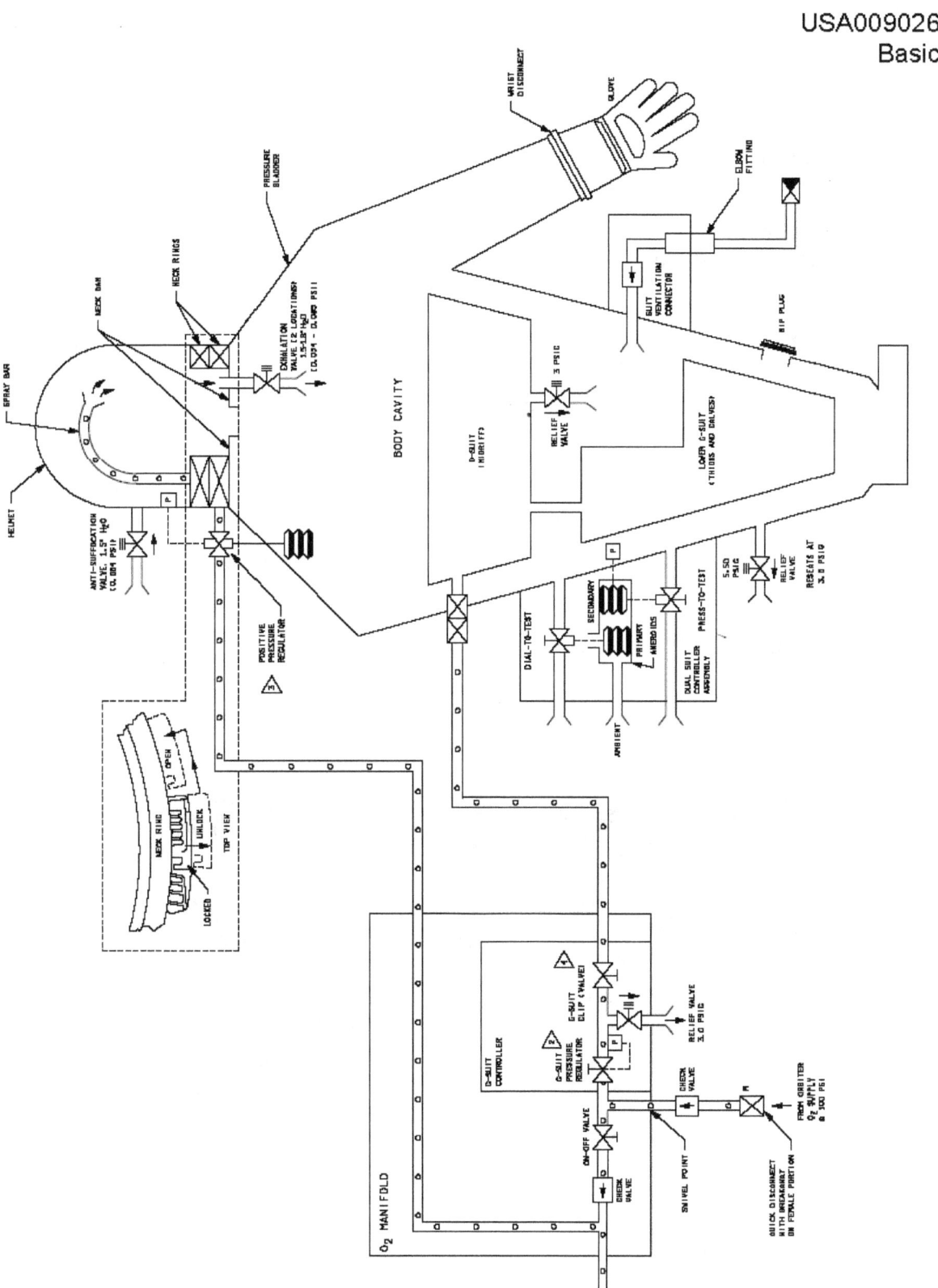

Figure 2-2. ACES oxygen flow and pressure system

2.2.4 Pressure System Components

The ACES, a full pressure suit, is a pressure vessel. The single pressure bladder covers all but a crewmember's head and hands.

The pressure on the crewmember's body is atmospheric and the pressure is being exerted directly and providing "an atmosphere of protection" indefinitely from a loss of cabin pressure to 100,000 ft.

The pressure bladder is constructed of a single nylon layer laminated to Gore-Tex and is restrained with an integral-like cover.

a. The single layer makes the ACES looser fitting, giving the crewmember greater freedom of movement and comfort.

b. The Gore-Tex is porous and "breathes" (helping the crewmember stay cooler), but seals when air pressure is applied.

Besides the pressure bladder, the ACES pressure system includes the following components:

2.2.4.1 Dual Suit Controller Characteristics/Function

Located on right side of suit chest (depicted below, with manual controls detailed).

Opens and closes to regulate suit pressure.

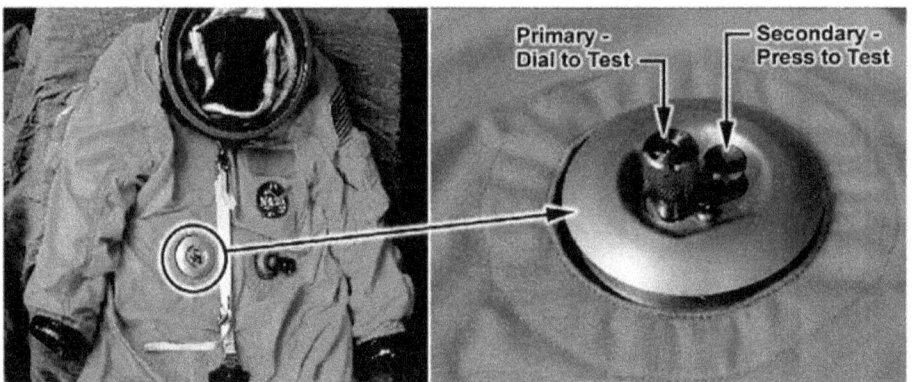

Figure 2-3. Dual suit controller

The controller includes dual (redundant) aneroids that operate in auto and test modes:

Auto Mode

In auto mode, the dual suit controller operates as follows:

a. The dual suit controller maintains suit pressure, based on ambient pressure conditions. It will maintain total pressure (ambient cabin pressure + suit pressure) of 3.67 psia. For example, an ambient pressure of 1.00 psi (~60,000 ft) would require a suit pressure of 2.67 psig. The valve assembly is normally open, preventing suit pressurization when the helmet visor is closed.

b. The primary aneroid assembly closes at 3.24 - 3.67 psia ambient pressure to maintain a 34,000 - 36,500 ft suit pressure altitude.

c. The secondary aneroid assembly closes at 3.05 - 3.48 psia ambient pressure to maintain a 35,000 - 38,000 ft suit pressure altitude.

Test Mode

The suit controller has the two following means of testing ACES pressurization:

a. <u>Dial-to-Test feature</u> – When the knurled knob in the center of the controller is turned clockwise approximately one rotation, it closes the primary aneroid assembly regardless of altitude. If the helmet visor is closed and locked, crewmember exhalation pressurizes the suit bladder to ~3.5 psi. Once inflated this way, the suit bladder remains pressurized (even if the visor is later opened) until the knob is turned one full turn counterclockwise.

b. <u>Press-to-Test button</u> – Also located near the center of the suit controller. If the helmet visor is closed and locked, depressing the button closes the secondary aneroid assembly and allows the suit bladder to inflate to 2.3 psi as long as the button remains depressed.

Note: <u>If you find your suit pressurizing unexpectedly</u>:

1. You may have inadvertently closed the Dial-to-Test knob (primary aneroid assembly)

or

2. You may be pressing in on the Press-to-Test button (secondary aneroid assembly)

2.2.4.2 Suit O_2 Regulator Characteristics/Function

During normal use, pressurized O_2 from the orbiter supply enters the helmet through the positive pressure regulator when the crewmember inhales. The positive pressure regulator maintains the pressure inside the helmet at 0.8-1.4 in of water (0.0289-0.0505 psi) above the pressure inside the suit.

Note: If breathing cavity integrity is compromised, the regulator will freeflow O_2 at a rate of 90 liters per minute.

2.2.4.3 Exhalation Valves Characteristics/Function

Two exhalation valves between the helmet cavity and the suit bladder will open if the pressure inside the helmet is 1.5-1.8 in of water (0.054-0.085 psi). This will happen during exhalation. Exhaled air passes into the suit bladder and out the dual suit controller on the right side of the chest.

2.2.4.4 ACES Relief Valve Characteristics/Function

The ACES relief valve is located in the right leg area. The valve opens when suit bladder pressure reaches 5.5 ± 0.2 psi. Once open, the valve stays open until pressure in the suit bladder is reduced to 3.5 ± 0.02 psi.

2.2.5 Anti-g Protection

Crewmembers wear a g-suit for entry that provides pressure to the lower body separate from the ACES. The applied pressure prevents blood from pooling in the lower extremities upon return to 1-g conditions after two or more days of microgravity.

Crewmembers wearing the ACES suit do not wear a g-suit for launch. The g-suit is stowed prelaunch, then donned and mated for entry.

The g-suit (Figure 2-4) has an abdominal bladder and leg bladders that do not cover the entire leg.

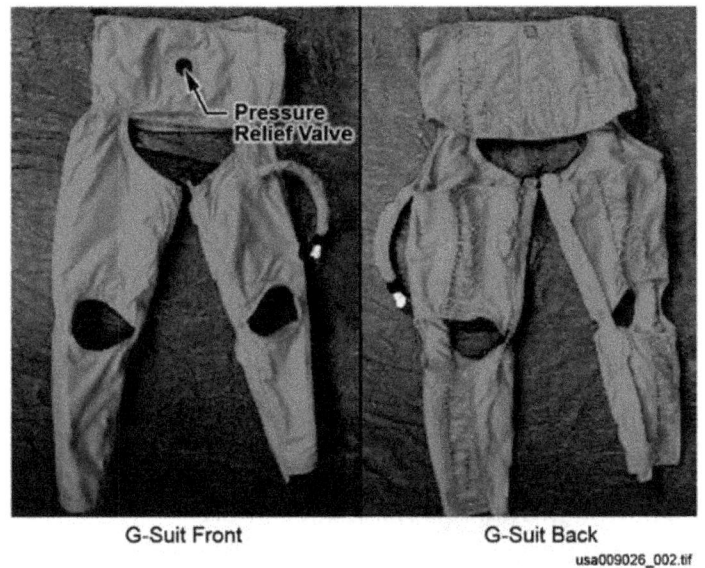

G-Suit Front   G-Suit Back

Figure 2-4. G-Suit

2.2.6 O₂ Manifold/G-Suit Controller Assembly

Inflation of the g-suit is controlled at the O_2 manifold/g-suit controller (Figure 2-5), which is a one-piece assembly located on the ACES upper left leg (Figure 2-1).

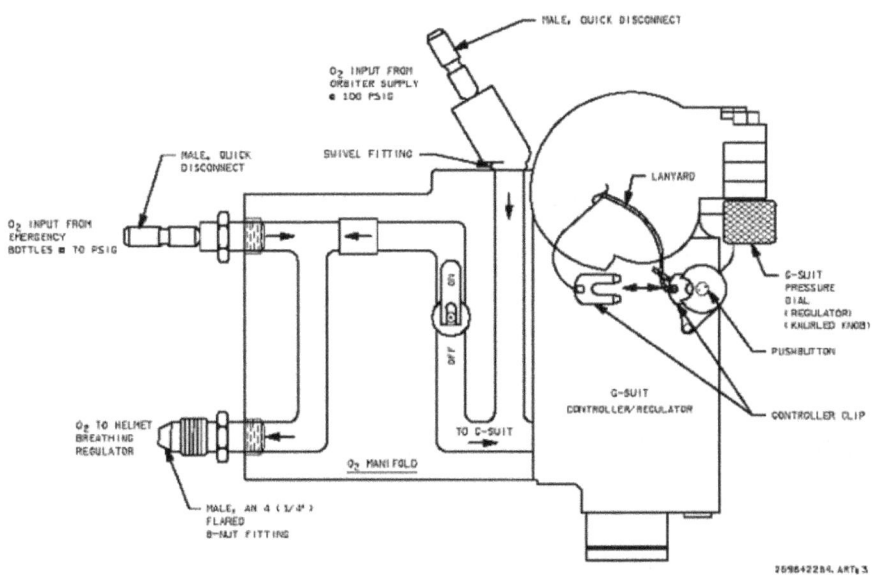

Figure 2-5. O₂ manifold/g-suit controller

The functional features of the O_2 manifold/g-suit controller assembly are as follows:

2.2.6.1 O_2 Hose Fittings Characteristics/Function

There are three O_2 hose fittings on the assembly:

a. A male swivel fitting connects to the orbiter O_2 supply hose. This fitting is provided with a quick disconnect.

b. A second male fitting, also a quick disconnect, connects the manifold with the emergency O_2 bottles in the parachute harness.

   Note: The emergency O_2 system does not provide g-suit protection.

c. A third (threaded) male fitting connects the manifold with the suit O_2 regulator for crew respiration.

2.2.6.2 G-suit Controller Characteristics/Function

The controller inflates the g-suit bladders when the manifold is connected to orbiter O_2 and the g-suit pressure dial (knurled knob) is turned clockwise (in the direction of the arrow).

The controller inflates the suit bladder 0.5 psi with each complete turn (indicated by a slight detent), to 2.5 psi max.

Note: Entry Checklist calls for 1.5 psi, or three complete turns.

Note: If you find the g-suit inflating unexpectedly, check that the g-suit pressure dial (knurled knob) is completely counterclockwise.

A one-way valve inside the manifold keeps the suit from inflating with emergency O_2, which is reserved for breathing.

2.2.6.3 G-suit Controller Clip Characteristics/Function

When the controller clip is pulled free, it:

a. Traps the current volume of O_2 inside the g-suit bladders and prevents any deflation, even if orbiter O_2 is disconnected from the manifold.

b. Prevents the g-suit from being inflated or deflated.

Reinstalling the clip allows the g-suit bladders to deflate or inflate.

2.2.6.4 G-suit Relief Valve Characteristics/Function

Located on the abdominal bladder of the g-suit. The valve operates as follows:

a. It vents pressure in excess of 3.0 psi out of the g-suit bladders.

b. Once open, the valve stays open until bladder pressure has been reduced to 2.5 psi.

2.2.7 ACES Gloves

The ACES gloves are shown in Figure 2-6.

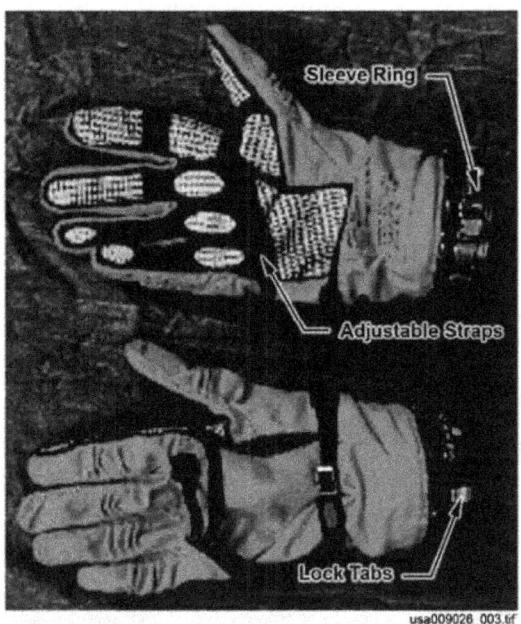

Figure 2-6. ACES gloves

The gloves, made of Nomex and silicone, have the following features:

a. Have adjustable straps that serve as pressure bars to reduce the rigidity of the gloves when the suit is pressurized.

b. Mate to the suit sleeve at the sleeve ring.

c. Have a locking ring with a directional arrow labeled "OPEN" and another arrow labeled "LOCK."

d. Alignment marks on the glove ring and on the sleeve ring show the proper alignment for mating the glove to the sleeve.

To mate glove to suit sleeve:

Step	Action
1	Align single mark on glove ring with triple mark on sleeve ring.
2	Twist glove until glove lock tabs snap into locking position.

To demate glove from suit sleeve:

Step	Action
1	Release and hold two locking tabs while twisting glove fully to mechanical stop.
2	Remove glove.

Each glove has a colored Velcro patch that helps identify individual crewmembers (see last paragraph).

2.2.8 Boots

The boots, made of leather, Nomex, and rubber, are equipped with lacing for fit adjustment. Each boot has a colored Velcro patch with a letter for crewmember identification during SAR operations (see next paragraph).

2.2.9 Crewmember Identification Patches

Colored Velcro patches with a letter help crewmembers identify personal equipment (suits, harness, boots, etc.) during on-orbit operations and to identify individual crewmembers during SAR operations (see Table 2-1 below). The helmet, parachute harness, and boots have this patch.

Table 2-1. Crewmember Color and Letter

Crew-member	Color	Letter	
CDR	Red	A	*Alpha*
PLT	Yellow	B	*Bravo*
MS1	Blue	C	*Charlie*
MS2	Green	D	*Delta*
MS3	Orange	E	*Echo*
MS4/PS1	Brown	F	*Foxtrot*
MS5/PS2	Purple	G	*Golf*
MS6/PS3/LON-CDR	Beige	H	*Hotel*
MS7/PS4/LON-PLT	Black	I	*India*
MS8/PSS/LON-MS1	Light gray	J	*Juliet*
MS3 DN	Beige	E	*Echo*
MS4 DN	Black	F	*Foxtrot*
MS5 DN	Light gray	G	*Golf*
LON-MS2	White	K	*Kilo*

2.3 HELMET, COMM CARRIER ASSEMBLY, HEADSET INTERFACE UNIT

2.3.1 Introduction

The helmet, shown in Figure 2-7, provides three critical functions for the crewmember:

a. A positive pressure-breathing environment.

b. An interface for communications between the Communications Carrier Assembly (CCA) and the Headset Interface Unit (HIU).

c. Protection to the crewmember's head.

The CCA, or "comm cap," contains all the equipment necessary for voice communication. The assembly is shown in Figure 2-9.

The HIU, depicted in Figure 2-10, provides volume control and Push-to-Talk (PTT) capability for both comm and intercom and preamplifies the microphone signal.

2.3.2 Helmet Features

The helmet features are shown in Figure 2-7 below.

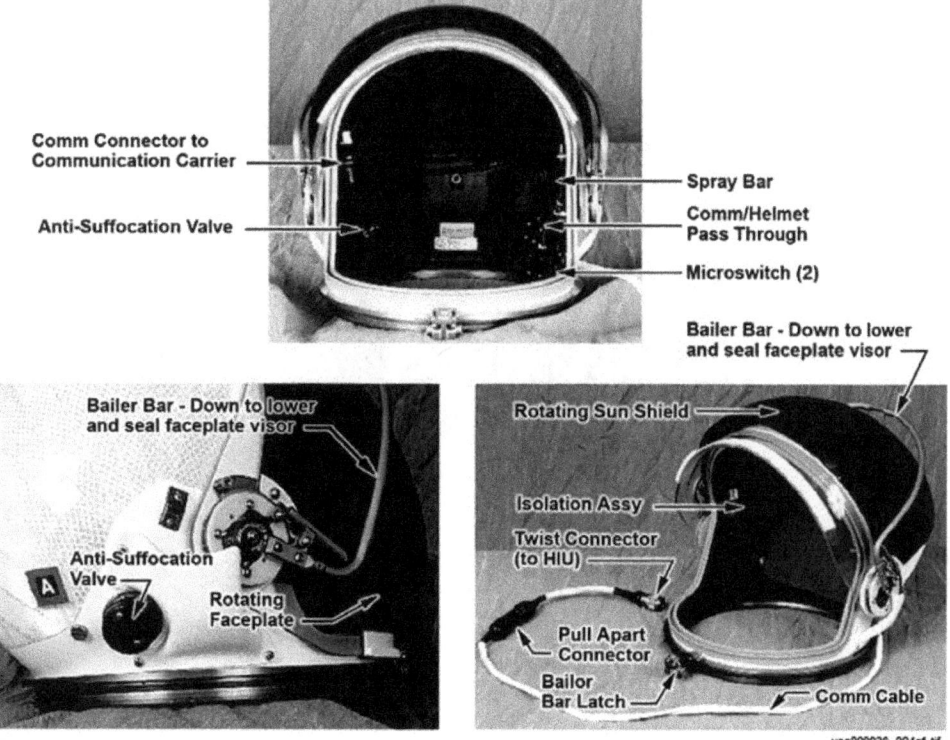

Figure 2-7. Helmet

The helmet attaches to the suit at the neck ring. When the helmet is locked onto the ring (Figure 2-8), the helmet can swivel from side to side.

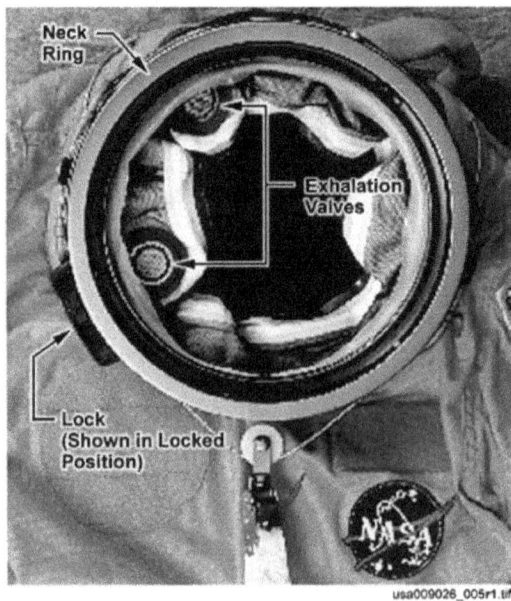

Figure 2-8. Neck ring

The neck ring lock has three positions:

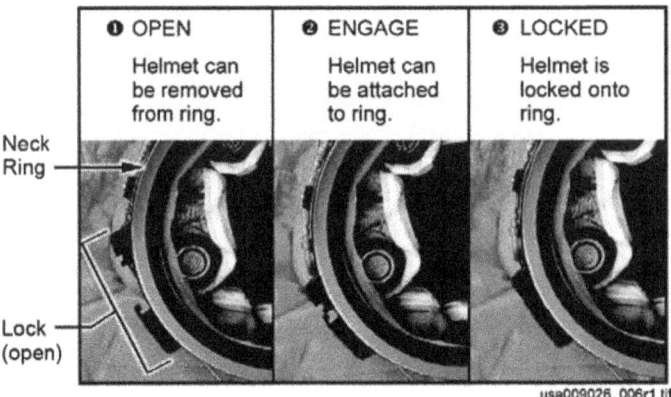

The helmet, which weighs ~7.5 lb, includes the following components:

2.3.2.1 Pressure Visor

A pressure visor that operates as follows:

a. To CLOSE, release neck dam tabs, grasp white tabs on rim of clear visor, and pull down.

b. To LOCK, pull bailer bar down into bailer bar latch.

c. To UNLOCK/RAISE:

    1. Push ribbed locking lever down

    2. Squeeze together two buttons on either side of bailer bar latch

    3. Manually raise visor

> **CAUTION**
> Do not close visor by pulling down on bailer bar. Doing so will damage the mechanism.

2.3.2.2 Rotating Sun Visor

A tinted sunshade is mounted to the outside of the helmet with a continuous friction hinge.

Can be lowered/raised independently of the clear visor.

2.3.2.3 Comm Cable/Comm Pigtail

The comm cable exiting the back of the helmet is equipped with a quick disconnect for egress/bailout. It connects to the comm pigtail, which in turn connects to the HIU.

An extra pigtail (connected to an extra HIU) is carried on the flight deck. Refer to flight specific crew worn carry on drawing for location.

2.3.2.4 Anti-Suffocation Valve

Located at the lower right rear of the helmet.

Allows ambient air outside the helmet to enter when:

a. Visor is closed.

b. Breathing O_2 is not available.

The anti-suffocation valve opens when pressure inside the helmet is 1.5 in. water (0.054 psi) less than the pressure outside the helmet (as happens when a crewmember inhales with the visor closed and no O_2 is being supplied).

Note: The anti-suffocation valve will allow water into the helmet if submerged.

2.3.3 CCA

The CCA contains all equipment necessary for voice communication. It is shown in Figure 2-9, alongside the helmet, with the comm connectors identified.

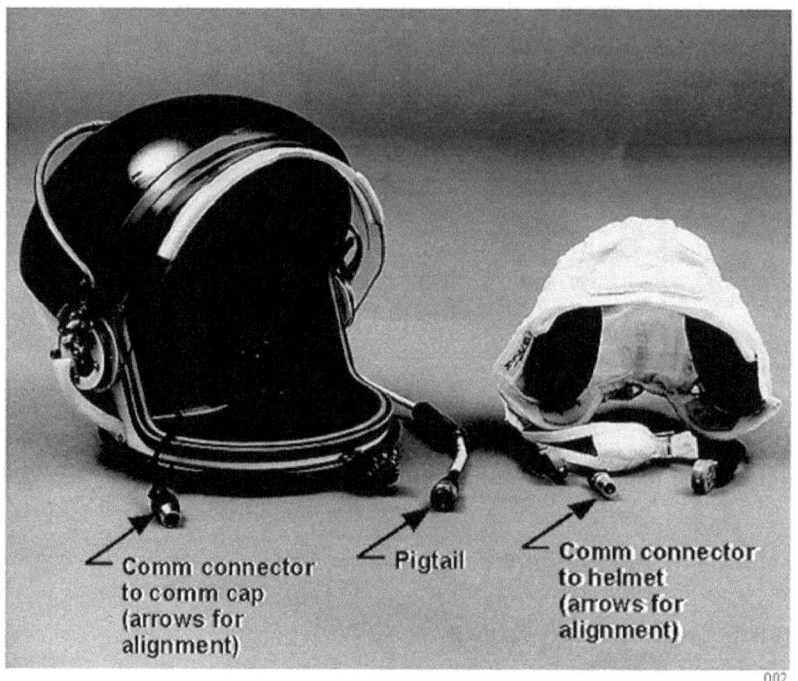

Figure 2-9. The CCA

2.3.4 CCA Components

CCA components include the following:

2.3.4.1 Microphones/Microswitches Characteristics/Function

The CCA has two independent (redundant) noise-canceling microphones.

When O_2 flows into the helmet, microswitches deactivate the microphones, auto-muting them.

Note: If the helmet does not have a good seal, O_2 will flow continuously, causing continuous auto-muting.

2.3.4.2 Earphones Characteristics/Function

The CCA has two independent (redundant) padded, earmuff-type earphones.

2.3.4.3 Connector Characteristics/Function

The comm cap interfaces with the helmet via a connector. See Figure 2-9.

2.3.5 HIU

The HIU (Figure 2-10) provides volume control and push-to-talk (PTT) capability and amplifies the microphone signal.

An extra HIU (connected to an extra pigtail) is carried on the flight deck. Refer to flight specific crew worn carry on drawing for location.

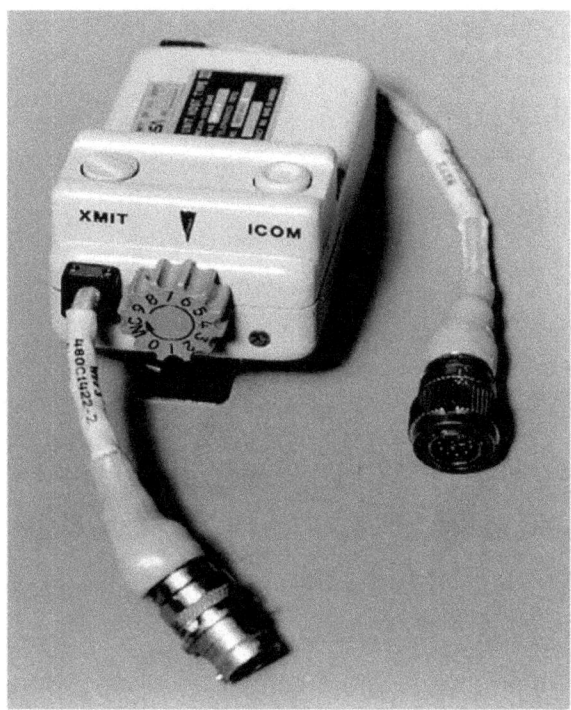

Figure 2-10. HIU

2.3.5.1 Volume Control

Provides volume control from "0" (min.) to "9" (max.) for earphones.

Fine-tunes the volume level selected by the crewmember at the Audio Terminal Unit (ATU).

2.3.5.2 "XMIT" PTT Button

Provides push-to-talk capability for transmitting on the Air-to-Air (A/A) and Air-to-Ground (A/G) loops.

The top surface of the button has a ridge (or "rib") across the diameter for easy identification by touch.

2.3.5.3 "ICOM" PTT Button

Provides push-to-talk capability for transmitting on the intercom loops.

The top surface of the button is concave ("like a corpuscle") for easy identification by touch.

2.3.5.4 Retainer Clip

Allows unit to be attached to crew clothing during on-orbit operations.

2.3.5.5 Preamplifier

Amplifies the signal from comm cap microphones.

2.4 COOLING SYSTEMS

2.4.1 Introduction

Each crewmember is provided with a cooling system. There are two models that utilize thermal electric modules for suit cooling: Individual Cooling Units (ICUs) and Thermal Electric Liquid Cooling Units (TELCUs), also known as "two-person cooling units."

2.4.2 TELCU/ICU Cooling

The two main components of the TELCU/ICU cooling system are:

a. The TELCU/ICU that cools the water flowing through the network of vinyl tubing in the Liquid Cooling Garment (LCG).

b. The two-piece LCG worn by crewmembers as an undergarment.

The TELCU/ICU cooling system is shown in Figure 2-11.

USA009026
Basic

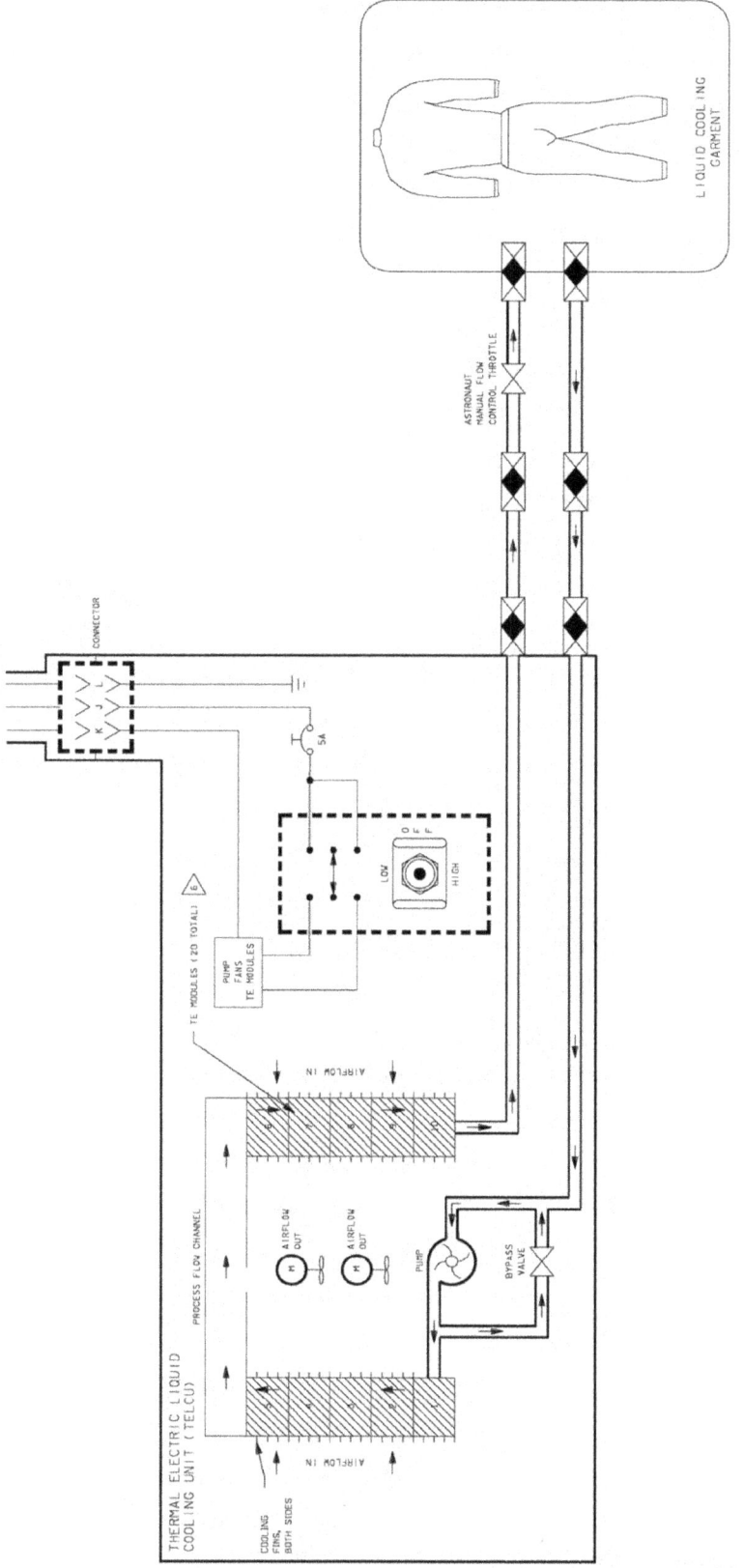

Figure 2-11. TELCU/ICU cooling system

The stages in the TELCU/ICU cooling system are as follows:

Stage	Action
1	Water from the liquid cooling garment enters the cooling side of the TELCU/ICU cooling core.
2	Two conductors are connected and a direct current is passed throughout the circuit. The temperature of one of the conductors decreases and the temperature of the other conductor increases. The cold conductor absorbs heat from its environment, and the hot conductor rejects heat to its environment. A pair of cooling fans aids the process.
3	The TELCU/ICU pump delivers the cooled water through the TELCU/ICU lines to the LCG lines via quick disconnects at the Bioinstrumentation Passthrough (BIP).
4	The cooled water circulates through the vinyl tubing network of the LCG worn by the crewmember.
5	The water becomes warmer during the heat exchange and returns to the TELCU/ICU, where the process starts over.

2.4.3 LCG

Crewmembers wear the two-piece LCG under the ACES. The LCG is used with the TELCU/ICU (previous subsection), which cools and pumps water through the LCG network of vinyl tubing to cool the wearer.

Both the LCG top and bottom have water hoses with quick disconnects at the BIP. Either garment part can be used with the TELCU/ICU. The two parts and the water hoses and fittings are shown in Figure 2-12 below.

It is a crew option to wear the LCG top without the LCG bottom and vice versa. When only the LCG top is worn, a jumper is required to complete the flow path.

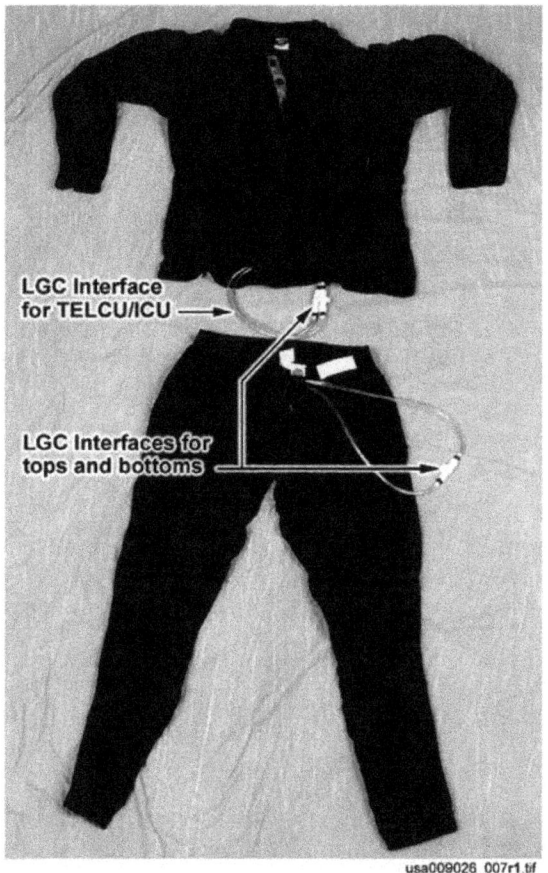

Figure 2-12. LCG

2.5 PARACHUTE HARNESS, PERSONAL PARACHUTE ASSEMBLY

2.5.1 Introduction

The parachute harness, which supports the crewmember's body weight, attaches to the Personal Parachute Assembly (PPA) containing the parachute system. Both the harness and the PPA include survival and rescue gear.

The harness and the PPA are discussed in separate subsections that follow. The rescue and survival gear associated with each system is described in Section 2.6.

2.5.2 Parachute Harness

The parachute harness has:

a. Straps that support the crewmember.

b. Fittings/snaps that attach the harness to the parachute.

c. Items that may be needed in an emergency:

   1. Emergency Oxygen System (EOS)

   2. Life Preserver Unit (LPU)

   3. Carabiner

   4. Emergency drinking water

These elements are shown in Figure 2-13 and on the next page. Harness elements are described in the table that follows (for the rescue and survival gear, refer to Section 2.6).

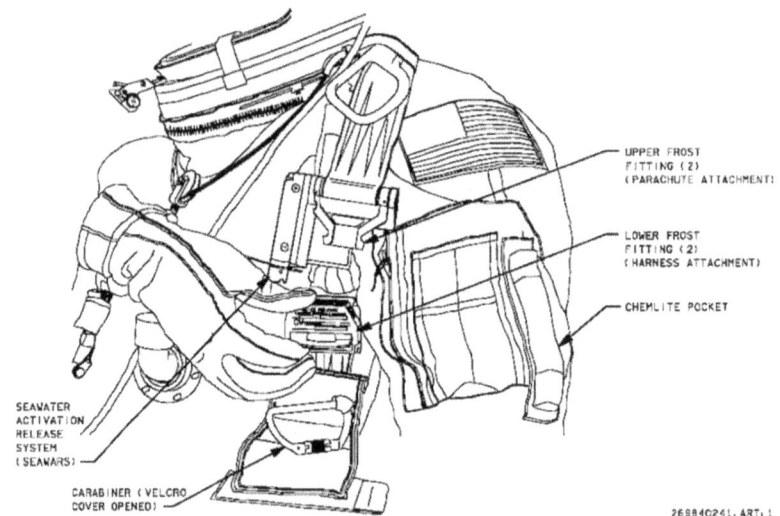

Figure 2-13. Parachute harness (upper portion)

The parachute harness elements include the following:

2.5.2.1   Support Straps

The system of nylon straps, depicted below, completely supports a crewmember's body weight during emergency egress, bailout, and rescue operations.

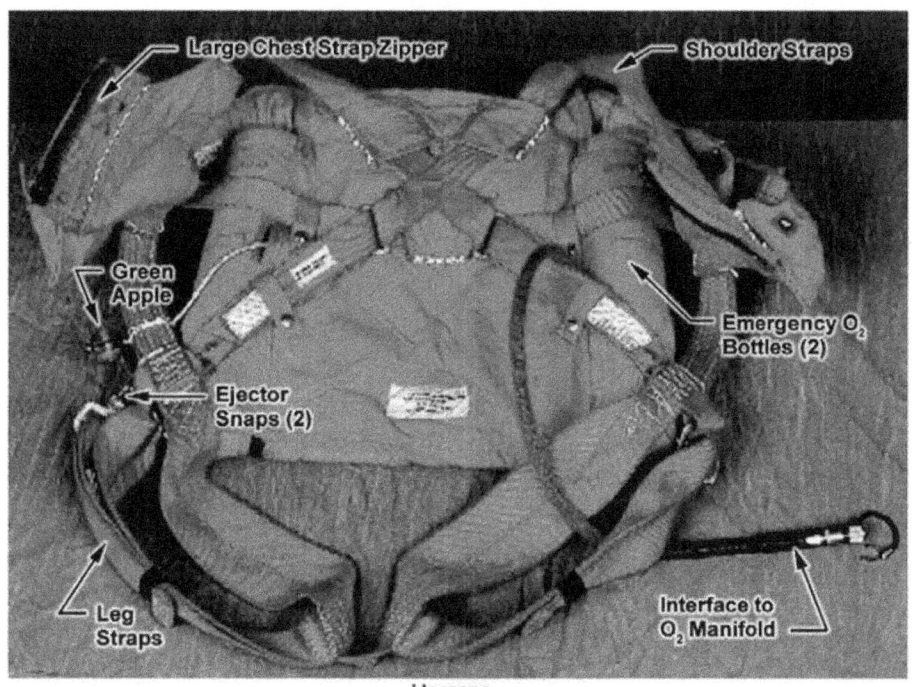

Harness
(open and ready for donning)

Figure 2-14. Harness

To don the harness:

Step	Procedure
1	Step through both leg loops.
2	Place arms through both sleeve openings.
3	Zip the vest zipper (under neck ring tiedown strap).
4	Connect and tighten the chest strap (under neck ring tiedown strap).

2.5.2.2 Parachute Attachment Points

The parachute attach points include:

Two Frost fittings (Figure 2-15)

Used to attach parachute risers to harness.

USA009026
Basic

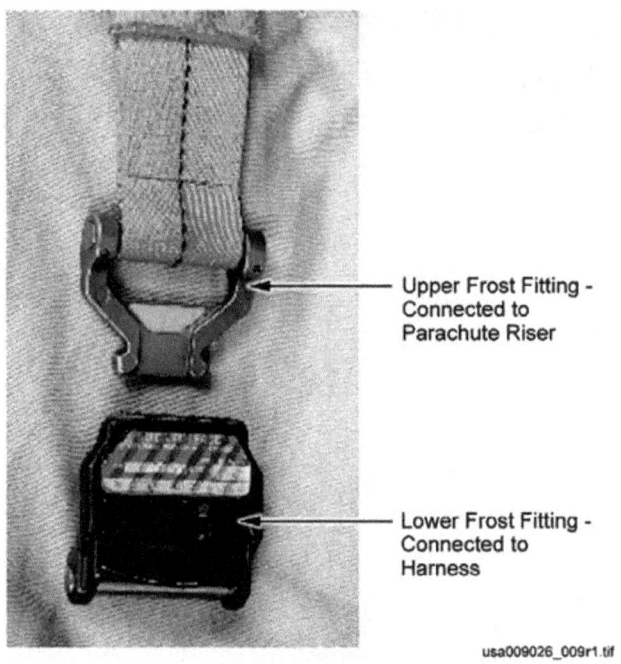

Figure 2-15. Frost fitting and riser

To attach:

Step	Procedure
1	Squeeze the top and bottom together on the harness attach fitting (female portion).
2	Insert parachute riser (male) fitting and snap into place.
3	Verify proper connection (no red mark seen in "mirror").

To release:

Step	Procedure
1	Squeeze the top and bottom together on the harness attach fitting (female portion).

Two ejector snaps (see illustration below)

a. Lower harness attach points connect the lower parachute attach fittings (triangular ring shape).

b. The ejector snaps have a soft dock position and a locked position (see below).

2-25

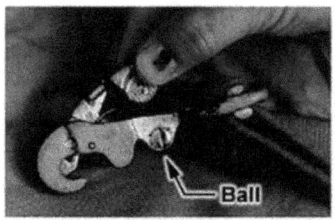
OPENED

SOFT DOCK

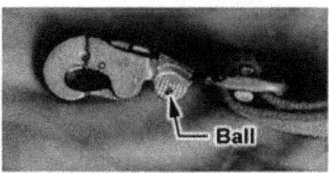
LOCKED

Ejector Snap Positions
(lower parachute attach fittings)

Figure 2-16. Ejector snaps

2.5.3 Parachute Assembly

The PPA, Figure 2-17, includes:

a. The parachutes, risers, and associated items.

b. The Search and Rescue Satellite-Aided Tracking (SARSAT) radio beacon.

c. A one-person life raft.

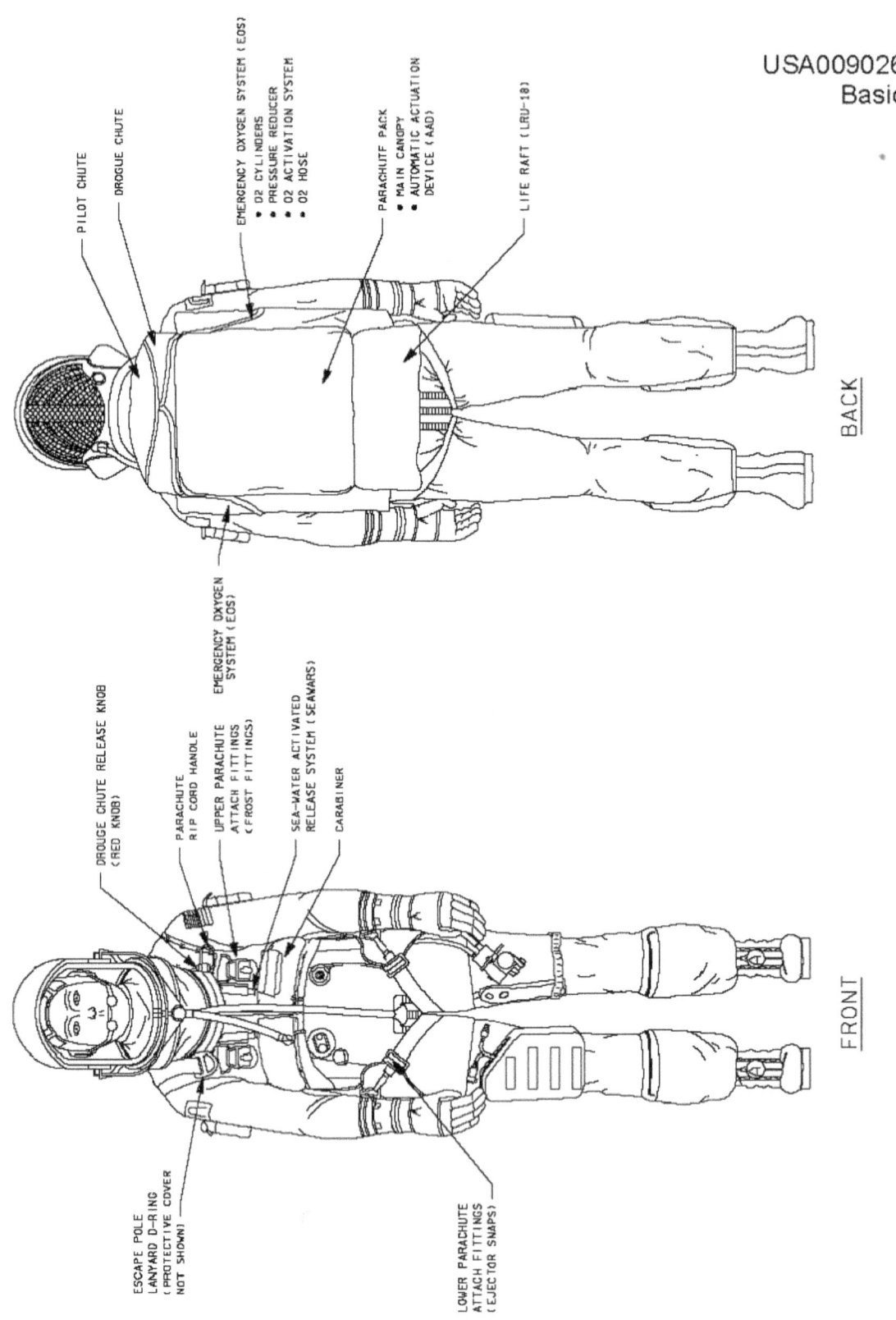

Figure 2-17. Personal parachute assembly

The PPA is:

a. Used as a seat back cushion.

b. Placed on the crew seats before ingress.

c. Connected during prelaunch to the parachute harness at four attach points.

The Table 2-2 below describes the PPA components (for the rescue/survival gear, refer to Section 2.6).

Table 2-2. PPA Components

Component	Description
Pole – Lanyard attach D-ring and bridle	The D-ring serves as the attachment point for the bailout lanyard located on the pole.
Pyrotechnic deployment system for the parachutes	Consists of: a. Three light-duty cutters for deployment of the pilot chute. b. Two heavy-duty cutters for bridle/lanyard separation. c. An Automatic Actuation Device (AAD) for the main canopy.
Parachute ripcord	The manual backup method of activating automatic deployment sequence. Located on left parachute riser.
Drogue chute release knob	Used only as a manual override to the AAD to deploy main canopy (if drogue chute already deployed).
Pilot chute	Has an 18-in. diameter canopy; pulls drogue chute from parachute pack.
Drogue chute	Has a 4.5-foot-diameter canopy; stabilizes crewmember and extracts main canopy from parachute pack.
Main canopy	A modified 26-foot-diameter circular military canopy with: a. Four anti-oscillation windows. b. A pyrotechnic reefing system to reduce the opening shock of the parachute on the crewmember.
SARSAT radio beacon	See Section 2.6
One-person life raft	
Pair of standard military Seawater-Activated Release Systems (SEAWARS)	One is attached to the front of each parachute riser. When immersed in salt water, the two SEAWARS automatically release the risers, separating the crewmember from the canopy. Note: In the unlikely event that the SEAWARS fails, squeeze top and bottom of Frost fittings to disconnect.
Raft inflation lanyard	Used to manually inflate and deploy the life raft.

2.5.4 Parachute Deployment

Figure 2-18, Figure 2-19, and Figure 2-20 on pages that follow illustrate respectively:

a. The automatic parachute deploy sequence.

b. The manual parachute deploy sequence.

c. The automatic parachute deploy sequence using the red knob.

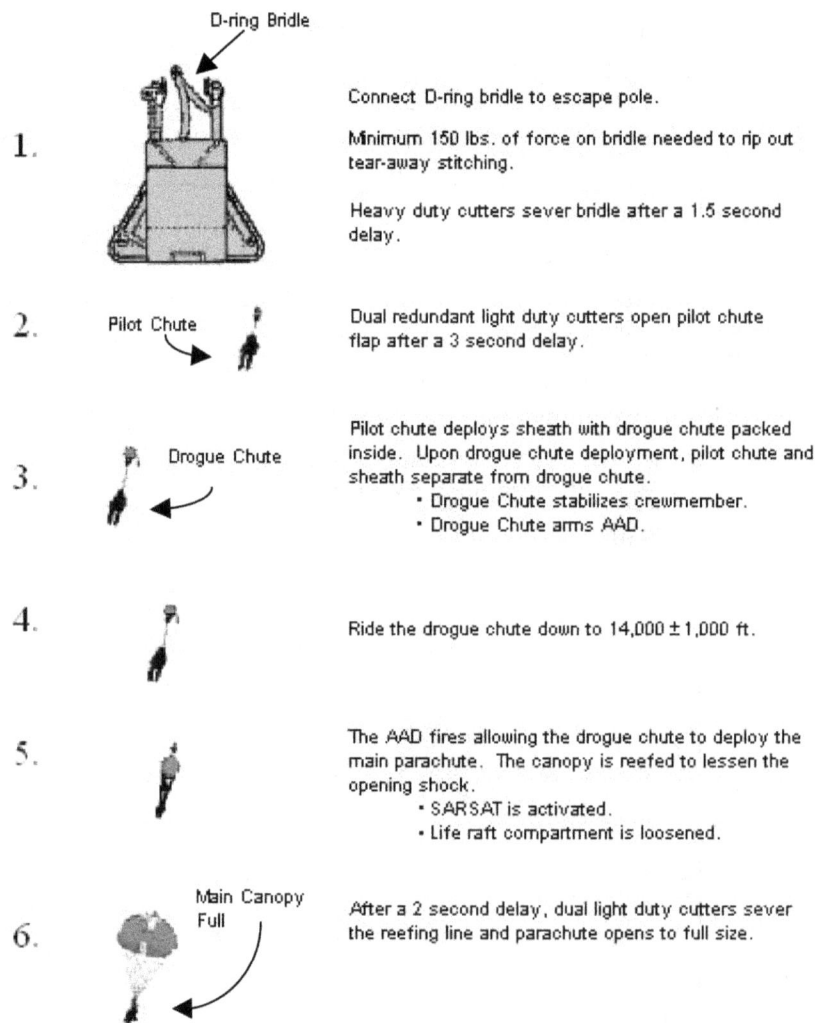

Figure 2-18. Automatic parachute deploy sequence

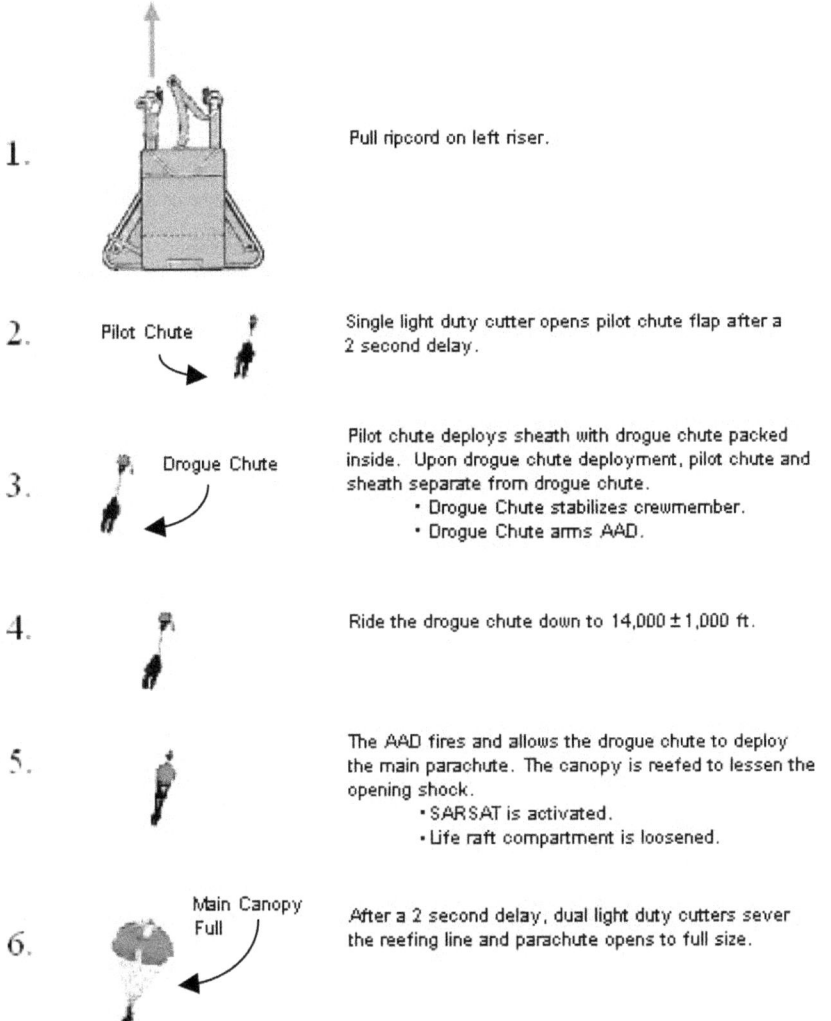

Figure 2-19. Manual parachute deploy sequence

USA009026
Basic

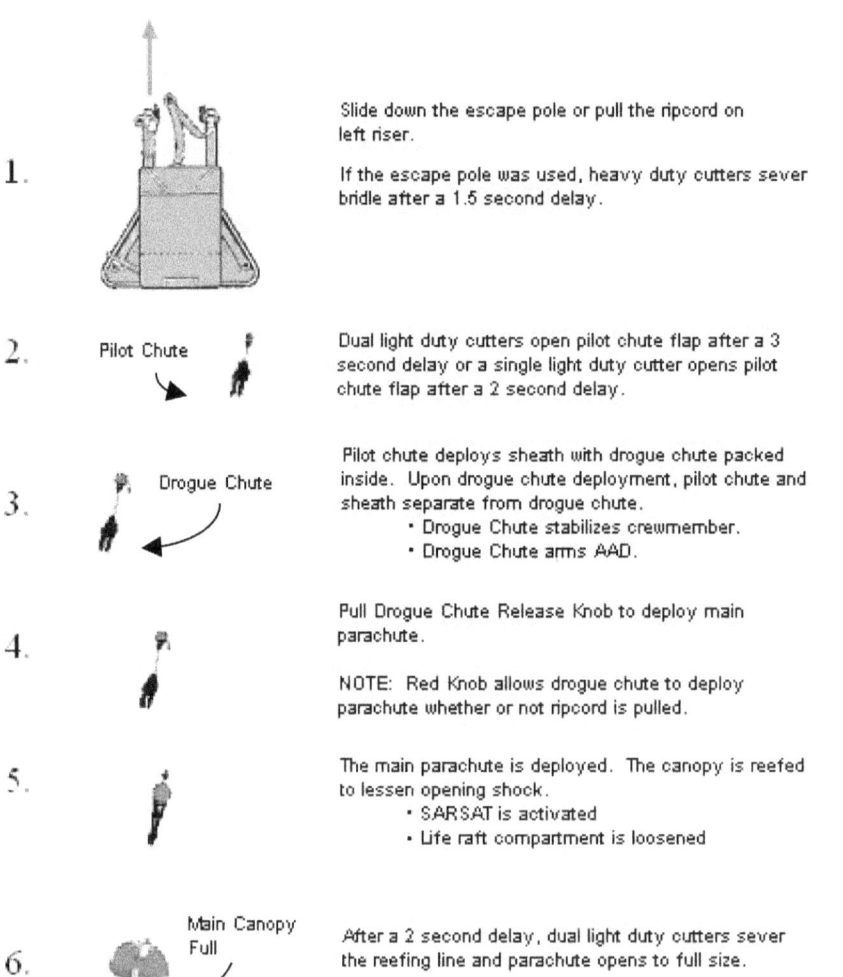

Figure 2-20. Automatic parachute deploy sequence using the red knob

2.6   RESCUE/SURVIVAL GEAR

2.6.1   Introduction

Crewmembers are provided an array of emergency rescue/survival gear. This gear is stowed in the following places:

a. Lower right and left suit leg pockets (there is also a shroudline cutter/knife located in a small pocket on the upper left suit leg).

b. Parachute harness.

c. Personal parachute assembly.

Subsections below depict and describe the gear by stowage location.

2.6.2 Gear in Left Leg Suit Pocket

The rescue/survival gear stowed in a packet (packet A) in the left leg suit pocket is illustrated in Figure 2-21 and described in the table that follows.

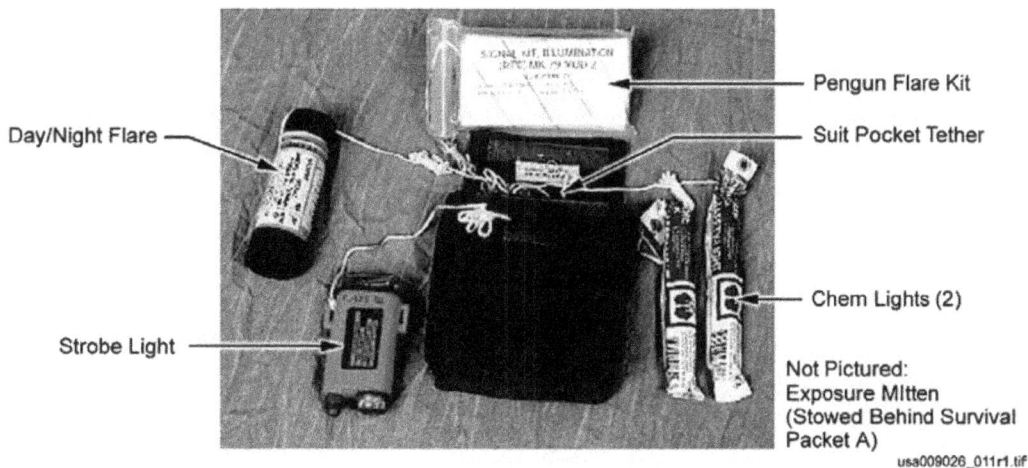

Figure 2-21. Rescue/survival gear, left leg suit pocket (packet A)

The section below describes the rescue/survival gear stowed in a packet (packet A) in the left leg suit pocket.

2.6.2.1 Pengun Flare Kit (MK-70)

Functional features:

a. Seven cartridges supplied.

b. Duration of each cartridge 4.5 sec.

c. Visible 50 mi.

Note: Use recommended only when visual contact made with SAR forces or SAR forces request use.

To operate:

Step	Procedure
1	With gun in "cocked" position, screw cartridge on gun.
2	Pull trigger knob and release (firing produces slight kick).

2.6.2.2 Two Chem Lights

Green 12-hr, Cyalume lightsticks.

To activate, bend to break glass vile and shake.

2.6.2.3 Day/Night Flare and Smoke Signal (MK-124)

Functional features:

a. Red smoke day; red flare night.

b. Duration: ~20 sec day/night.

c. Visibility: 7 mi day; 30 mi night.

d. Use gloves (flare and smoke generate heat).

Note: Use only when visual contact established with SAR forces or as directed by SAR forces.

To operate:

Step	Procedure
1	Remove cover.
2	Holding flare away from raft, pull trigger out then down to fire. Note: If smoke signal begins to flare, quickly dip in water.

2.6.2.4 Strobe Light

a. On/off switch.

b. Battery life 6 to 9 hr continuous operation; 18 hr intermittent.

c. Visible 50 mi.

2.6.2.5 Exposure Mitten

Other mitten is in right leg pocket.

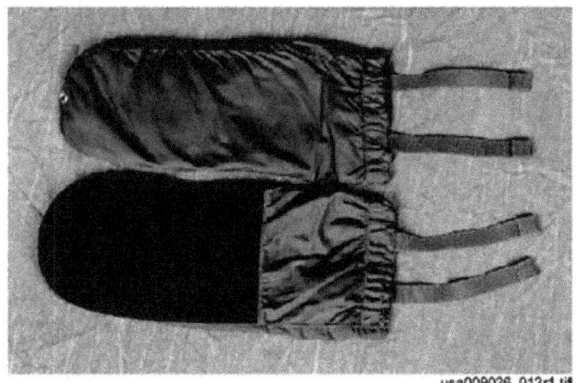

Figure 2-22. Exposure mittens

2.6.3 Gear in Right Leg Suit Pocket

The rescue/survival gear stowed in a packet (packet B) in the right leg suit pocket is illustrated in Figure 2-23 below and described in the table that follows.

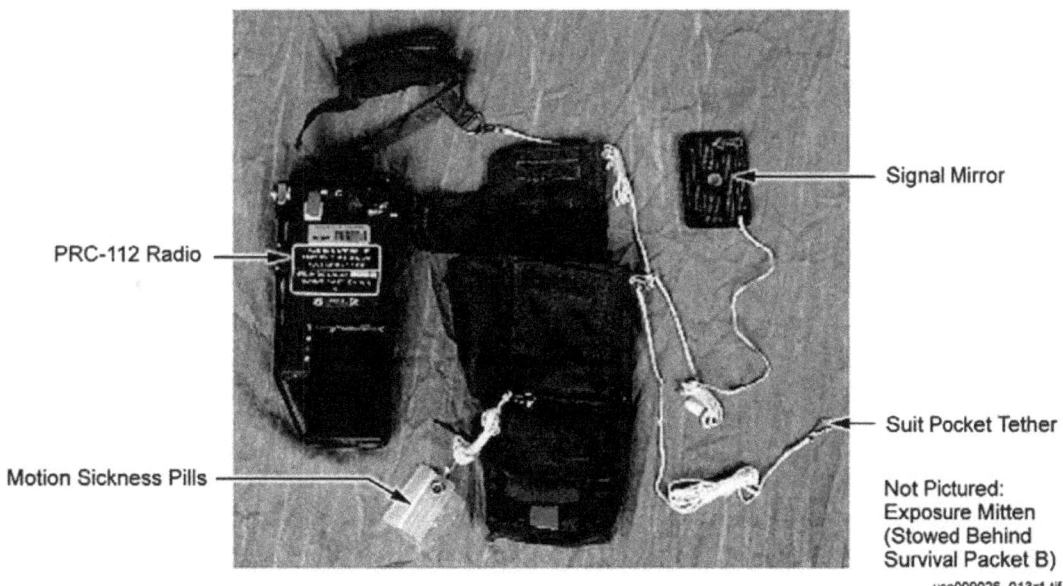

Figure 2-23. Rescue/survival gear, right leg suit pocket (packet B)

The section below describes the rescue/survival gear stowed in a packet (packet B) in the right leg suit pocket:

2.6.3.1   Survival Radio

The radio, a PRC-112, has:

a.   Transmit/receive capability.

b.   Continuous beacon capability.

Battery life is 24 hr (duty cycle of 10% transmit and 90% receive).

An earphone and spare antenna are in a packet attached to the radio.

To operate the radio, follow the decal on the radio back. The decal procedures are shown below:

PRC-112 Radio Instructions

1. Extend SARSAT telescoping antenna, then disconnect wire antenna
2. Continue SARSAT beaconing
3. Standby for radio contact with SAR forces
4. Follow instructions from SAR forces
5. Return to 282.8 if contact with SAR forces lost
6. Verify SARSAT beacon tones with PRC-112 by momentarily listening to 243.0 (voice) and 121.5 (voice)

Additional Information

121.5 BCN	International emergency frequency (VHF/beacon)
121.5	International emergency frequency (VHF/voice)
243.0 BCN	Military emergency frequency (UHF/beacon)
243.0	Military emergency frequency (UHF/voice), DME supported
282.8	Primary SAR operations frequency (UHF/voice), DME supported
A	Secondary SAR operations (259.7) frequency (shuttle A/G) (UHF/voice), DME supported
B	Backup SAR operations (236.0) frequency (UHF/voice), DME supported

2.6.3.2 Signal Mirror

Visible 40 mi. Instructions for use displayed on back.

2.6.3.3 Two motion Sickness Pills

0.4 mg Scopolamine, 5 mg Dexedrine in each pill.

2.6.3.4 Exposure Mitten

Other mitten is in left leg pocket.

2.6.4 Shroudline Cutter/Knife

A shroudline cutter/knife is located in a small pocket on the upper left leg of the suit.

2.6.5 Gear, Parachute Harness

The parachute harness (Figure 2-13) includes the following emergency gear:

2.6.5.1 Life Preserver Unit

The LPU, attached to the harness, has the following features:

a. Horseshoe-shaped bladder that fits under the crewmember's arms and around back.

b. Designed to keep an unconscious crewmember's head out of the water.

c. Inflates with CO_2 using two FLU-8 devices. FLU-8 are water activated or can be manually activated.

2.6.5.2 Emergency Oxygen System

Consists of two bottles of oxygen for use during emergency egress, which:

a. Are attached to the back of the harness.

b. Contain 120 in^3 O_2 at 3000 psi (381 liters usable O_2).

c. Delivers O_2, via reducer valves, at 70 psia.

d. At moderate level of exertion (38 liters/min of use), EOS will last 10 min at sea level and longer at high altitudes.

The figure below gives an overview of the EOS and the detail of the O_2 regulator/valve:

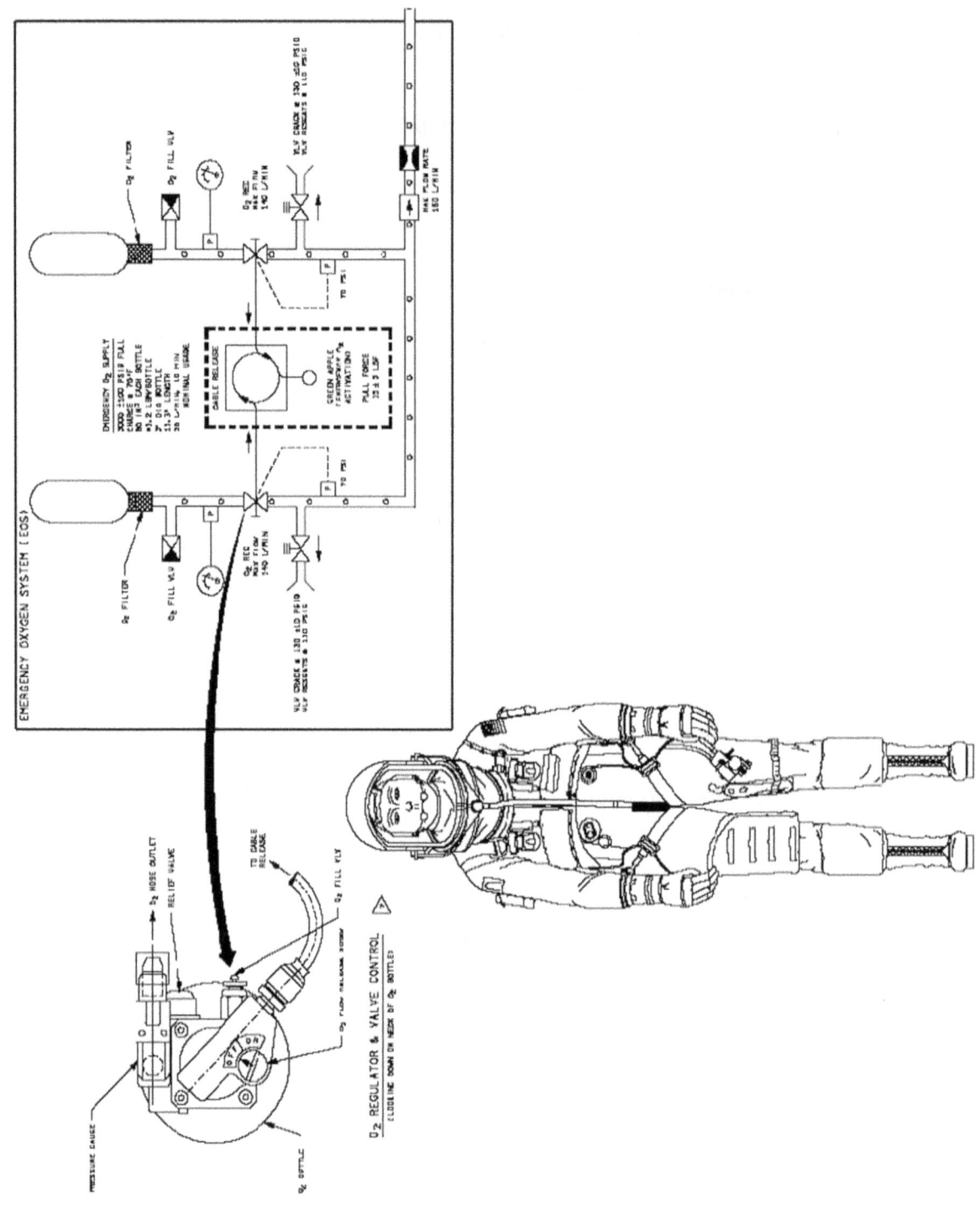

Figure 2-24. EOS and O₂ Regulator Valve

To activate the EOS:

Step	Procedure
1	Lower and lock helmet visor. Note: If helmet visor is opened, O_2 will flow continuously into helmet. If orbiter O_2 supply is not activated, EOS O_2 will flow until bottles are depleted.

2	Pull EOS activation lanyard ("green apple") on lower right front of harness.
	Note: There are two "clicks":
	  1. The first disconnects the green apple from the snap keeper.
	  2. The second initiates O_2 flow.

2.6.5.3 Carabiner

The carabiner (Figure 2-13), attached to right side of harness above the LPU, can be used to hoist a crewmember.

2.6.5.4 Emergency Drinking Water

The water pack containing emergency drinking water:

a. Is Velcroed to the rear of the harness between the EOS bottles.

b. Is tethered to the harness by a 2-foot tether.

c. Has 2 liters of water contained in 16 packets.

2.6.6 Gear in PPA/Life Raft

The rescue/survival gear stowed in the life raft, which is contained in the PPA, is illustrated in Figure 2-25 below and described in the following section.

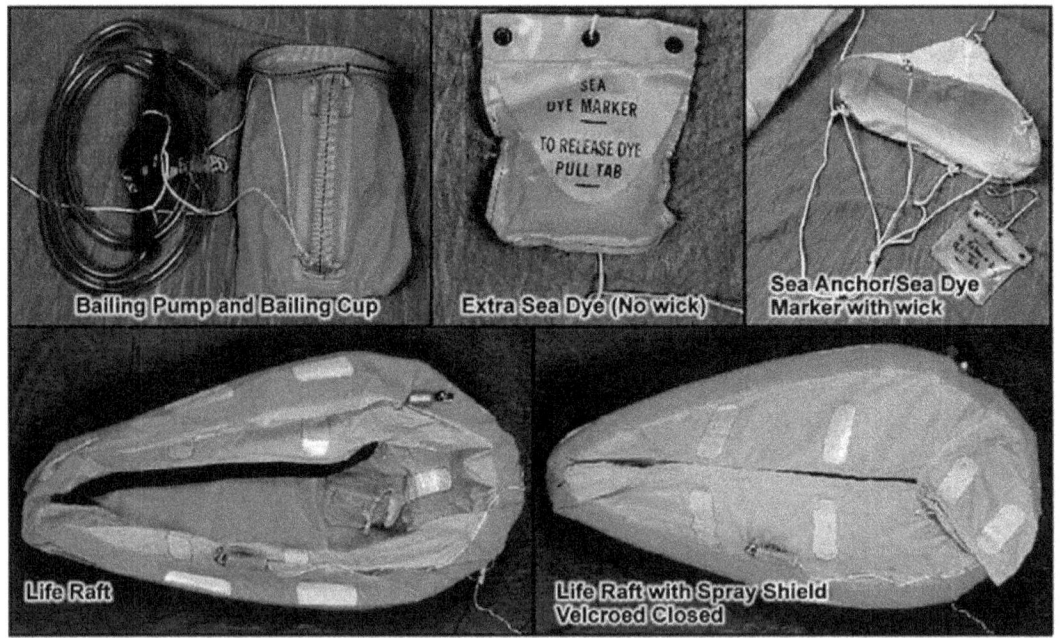

Figure 2-25. Rescue/survival gear, PPA/life raft

2.6.6.1 One-Person Survival Raft

a. Has eight rows of inflatable tubes:

   1. <u>Normally</u>, the tubes are inflated by two CO_2 cartridges as follows:

      (a) On entering the water, one of the CO_2 cartridges (FLU-9) automatically activates, inflating the second, third, and fourth tubes, thus giving the raft some shape and flotation. (The CO_2 cartridge can also be activated manually.)

      (b) After climbing in the raft, the crewmember manually activates the other CO_2 cartridge (FLU-1) to inflate the remaining five raft tubes.

   2. <u>If necessary</u>, the tubes can be inflated by using the two oral inflation tubes provided inside the raft.

   Note: Raft material is slightly permeable to CO_2.

   After several hours the raft will need to be inflated orally.

b. Is equipped with two spray shields. These shields are orange flaps that can be Velcroed closed.

c. Has reflective tape on sides and spray shield.

d. Is tethered to PPA remnant (the part that remains on crewmember) by 12-foot Kevlar tether.

e. Has bailing cup, bailing pump, and additional sea dye marker tethered inside the raft on the right side.

2.6.6.2 SARSAT Personal Locator Beacon

a. Flexible antenna loops around the back of the spray shield.

b. Extendable, rigid antenna manually deployed for better signal transmission.

c. On/off switch, battery life approximately 24 hours.

d. Beacon automatically activates upon parachute deployment, transmitting on the following frequencies:

121.5	International emergency
243.0	Military emergency
406.0	Satellite SAR

2.6.6.3 Sea Anchor/Sea Dye Marker

a. Tethered to outside of raft by a 12-foot tether.

b. The anchor is usually allowed to be untied by wave action.

c. Sea dye package that is designed to automatically wick dye when placed in the water.

d. Sea dye produces a green fluorescent slick highly visible by SAR forces.

Note: The dye dissipates quicker in rough conditions.

2.6.6.4 Additional Packet of Sea Dye

a. Tethered to inside of raft on right side.

b. Packet is sealed without wick.

c. Open packet when requested by SAR forces.

2.6.6.5 Bailing Cup, Bailing Pump

a. Bailing cup tethered to the inside of the raft on the right side.

b. Bailing pump is tethered to the bailing cup and stowed inside bailing cup.

c. The pump has directional flow as shown below.

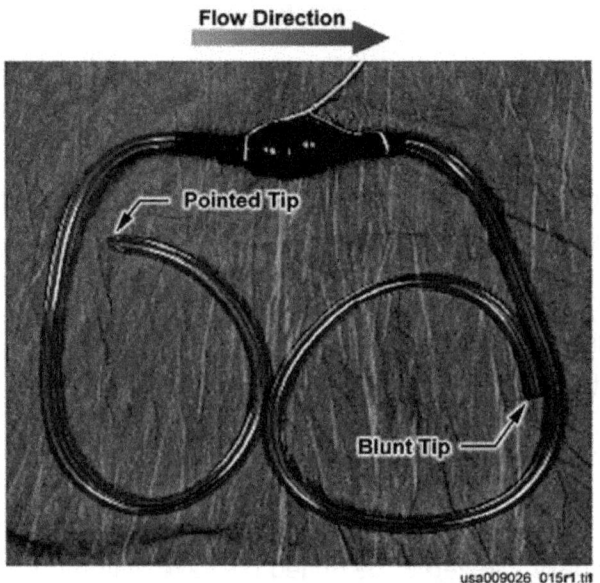

Figure 2-26. Bailing pump

3.0 ORBITER HARDWARE

3.1 OVERVIEW

3.1.1 Introduction

Crew escape hardware has been built into the orbiter that facilitates quick and safe crew egress/escape in an emergency. This hardware includes:

a. Crew seats.

b. Emergency egress net (trampoline), a platform for prelaunch egress.

c. The escape pole, for in-flight bailout.

d. The escape slide, for postlanding egress from the side hatch.

e. The Window 8 escape panel, secondary postlanding egress path in the event of total side hatch failure.

f. The descent control devices (Sky Genie), for postlanding egress in the event of complete slide failure.

g. The side hatch pyrotechnic jettison system and cabin vent system.

3.1.2 Purpose

This chapter describes the components, operational features, and operation of orbiter crew escape hardware.

3.1.3 Objectives

After completing this chapter, you will be able to identify:

a. Components of the crew seats

b. Components of the emergency egress net (trampoline).

c. Components of the escape pole and the steps in deploying and using the pole for bailout.

d. Components of the emergency egress slide system and the steps in using the system with:

    1. The side hatch on (hatch opened normally).

    2. The side hatch off (hatch jettisoned).

e. Components of the Window 8 escape panel and the steps in opening it.

f. Components of the descent control device (Sky Genie) and the steps in using it postlanding to descend from the orbiter through:

   1. The side hatch (primary route).

   2. The Window 8 escape panel (secondary route).

g. Components of the side hatch and the steps in opening the hatch normally and by jettisoning.

In This Chapter

Sections describe crew escape orbiter hardware as follows:

Section	Page
Crew Seats	3-2
Emergency Egress Net (Trampoline)	3-7
Side Hatch	3-12
Window 8 Escape Panel	3-18
Descent Control Device (Sky Genie)	3-27
Escape Pole	3-31
Emergency Egress Slide System	3-34

3.2　CREW SEATS

3.2.1　Description

Two types of seats are provided for crew positioning and restraint during all phases of flight. Each seat accommodates a fully suited crewmember, cooling unit, and FDF stowage bag. The seats can also secure oxygen hoses, communications, and power cables for launch and entry.

Figure 3-1. CDR and PLT seats

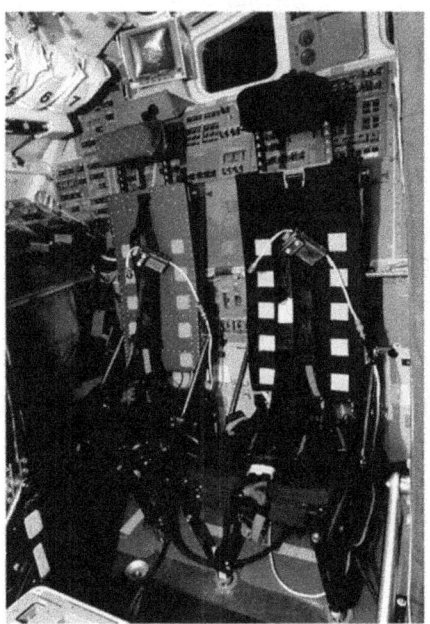

Figure 3-2. Mission specialist and payload specialist seats

Crewmembers are secured to the seat by a five-point harness. The shoulder and leg straps connect to a rotary buckle permanently mounted to the crotch strap. The straps are released by rotating the buckle either clockwise or counterclockwise.

Figure 3-3. 5-point Restraint

The commander and pilot seats are permanently installed on the orbiter flight deck. They are electrically powered for horizontal and vertical adjustment and equipped with mounting provisions for Rotational Hand Controllers (RHCs).

Figure 3-4. Horizontal and vertical adjustment

Mission Specialist (MS) seats are equipped with quick disconnect fittings that interface with mounting studs on the floors of the flight deck and middeck. The MS seats are removed, folded, and stowed for on-orbit activities.

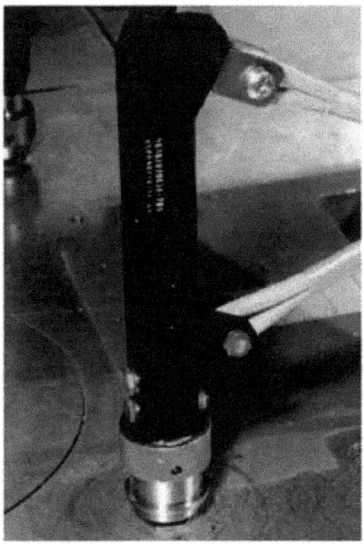

Figure 3-5. Quick disconnect fittings

All MS (PS) seats can be fitted with a emergency egress step for use when egressing the vehicle via the escape panel on Window 8. The step can be extended or stowed for launch and entry.

Figure 3-6. Emergency egress step

The MS seats can also be used with the Recumbent Seat Kit (RSK) for returning ISS crewmembers. The RSK consists of three sets of frames used to connect the specialists' seat back to the middeck floor.

Figure 3-7. Recumbent seat kit frames

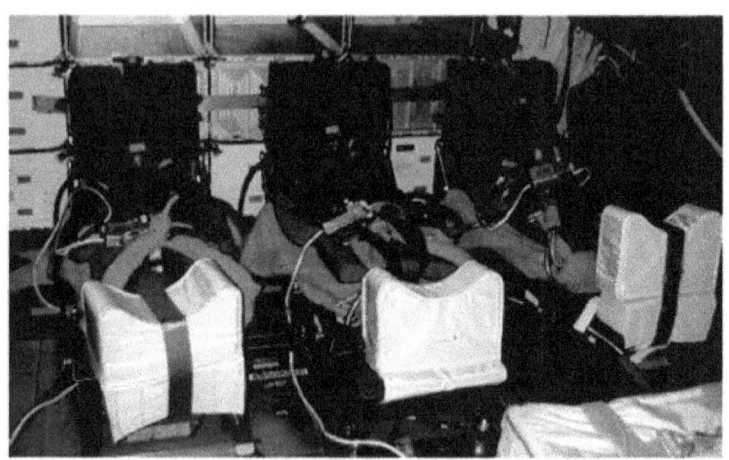

Figure 3-8. MS (PS) seats installed on RSK

3.2.2 Seat Functions and Features

3.2.2.1 Seat Back Angle Adjustment Operation

1. Pull aft on the lock lever located on right side of seat

2. Reposition back angle

3. Release lock lever

Note: All seat backs have two positions: 2° forward (launch position) and 10° aft (entry position). The recumbent seats are in the 10° aft position for entry.

3.2.2.2 Backrest Fold Down Operation

1. Pull up on T-handle located on seat back rear

2. Lower seatback

3. T-handle will snap into position

3.2.2.3 Headrest Height Adjustment Operation

1. Pull headrest height adjustment lock lever located on left side of seat back rear

2. Adjust headrest height

3. Release lock lever

3.2.2.4 Inertia Reel Lock/Unlock Operation

1. Push lever located on left side of seat forward to lock shoulder harness in place

2. Push lever aft to release

Note: Inertia reel locks automatically if rapid forward acceleration is experienced, even if inertia reel lever is in the unlocked position. To unlock inertial reel, cycle inertial reel lever forward and aft.

3.2.2.5 Adjustable Shoulder Harness/Lap Belt Operation

1. To connect, insert ends of the shoulder harness straps and lap belt straps into the connector

2. To disconnect, twist the connector rotary buckle either direction; all ends disconnect

3.3 EMERGENCY EGRESS NET (TRAMPOLINE)

3.3.1 Description

The Emergency Egress net, commonly referred to as the trampoline, serves as a platform for crew emergency egress when the orbiter is oriented vertically on the launch pad. With the removal of the internal airlock, a large volume was created behind seats 6 and 7 on the middeck. To allow quick and safe crew egress during an emergency, the trampoline is installed prelaunch by ground crews.

3.3.2 Trampoline Attachment

The trampoline is installed in the middeck, parallel to the aft bulkhead. Its span is from the deck to the ceiling and from the interdeck access ladder to the aft lockers.

Figure 3-9. Emergency egress net (Trampoline)

The trampoline attaches to both the ceiling and the floor via six double-acting pins through six fittings.

Figure 3-10. Trampoline attachment points – ceiling

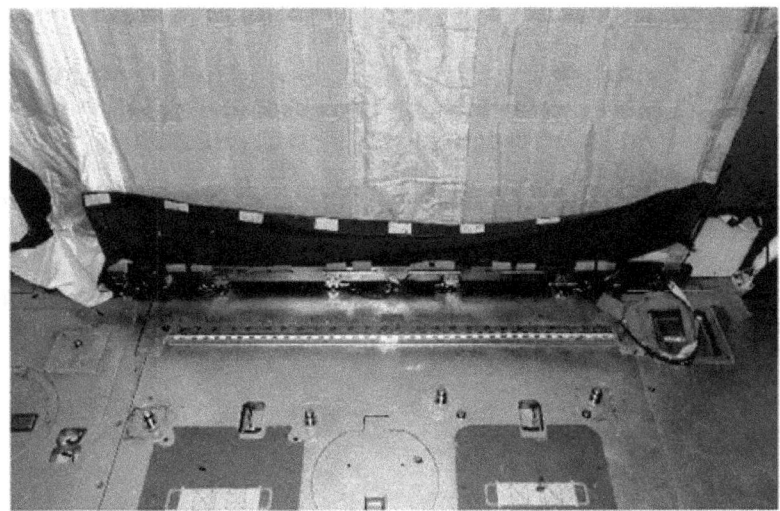

Figure 3-11. Trampoline attachment points – deck

Three straps are used to secure the trampoline across the edge of the egress platform, which is attached to the interdeck access.

The straps are wrapped across the front of the egress platform and attach to the inside of the ladder's forward leg post via "pull-the-dot" snaps.

Two ratchet assemblies at the port and starboard ceiling edges are rotated clockwise to tighten the trampoline fabric.

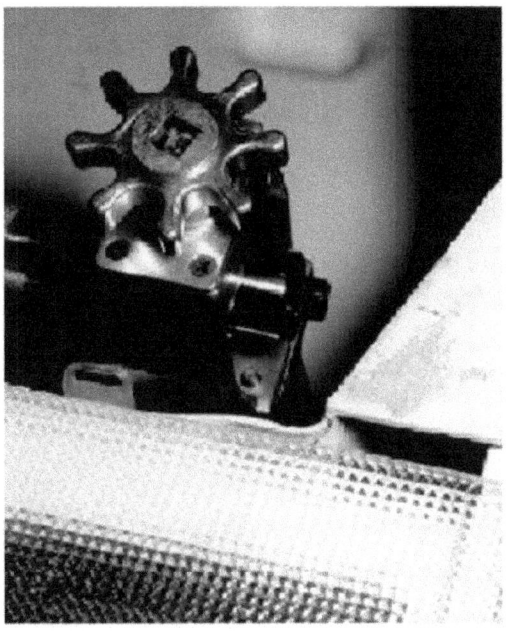

Figure 3-12. Ratchet assembly

3.3.3 Closeout Nets

Four closeout nets are installed around the trampoline to keep items from falling to the aft bulkhead. The brown mesh closeouts and are located at the ceiling, deck, and port side of the trampoline and are attached to crew module structure using Velcro.

The starboard side of the trampoline has a white Nomex extension that also acts as a closeout. This extension is attached to the aft middeck lockers with snap hooks.

These closeouts are released from module structure and rolled up inside the trampoline during stowage.

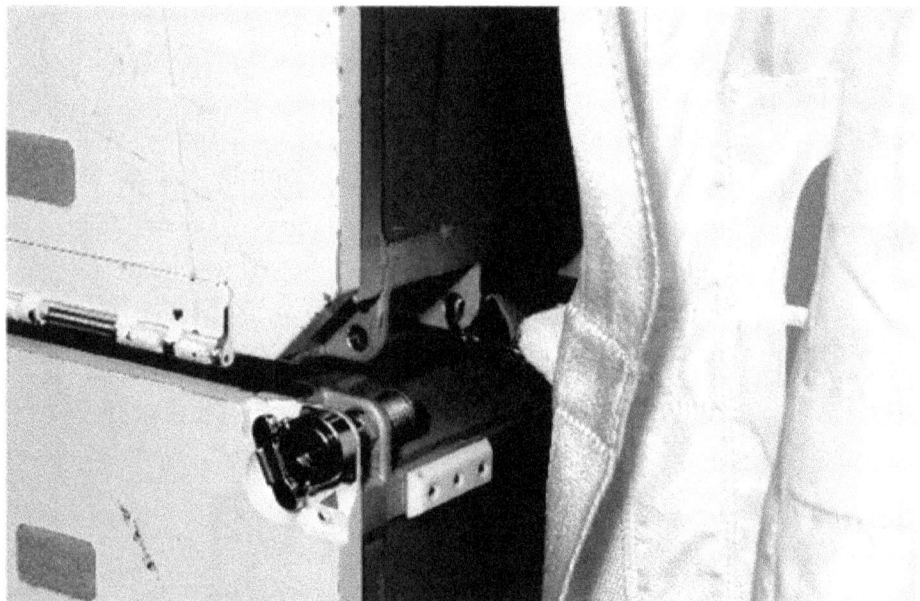

Figure 3-13. Nomex extension and snap hooks

3.3.4 Trampoline Stowage

On orbit, the crew releases the tension and disconnects the trampoline from the floor and rolls it up on the middeck ceiling for the duration of the flight.

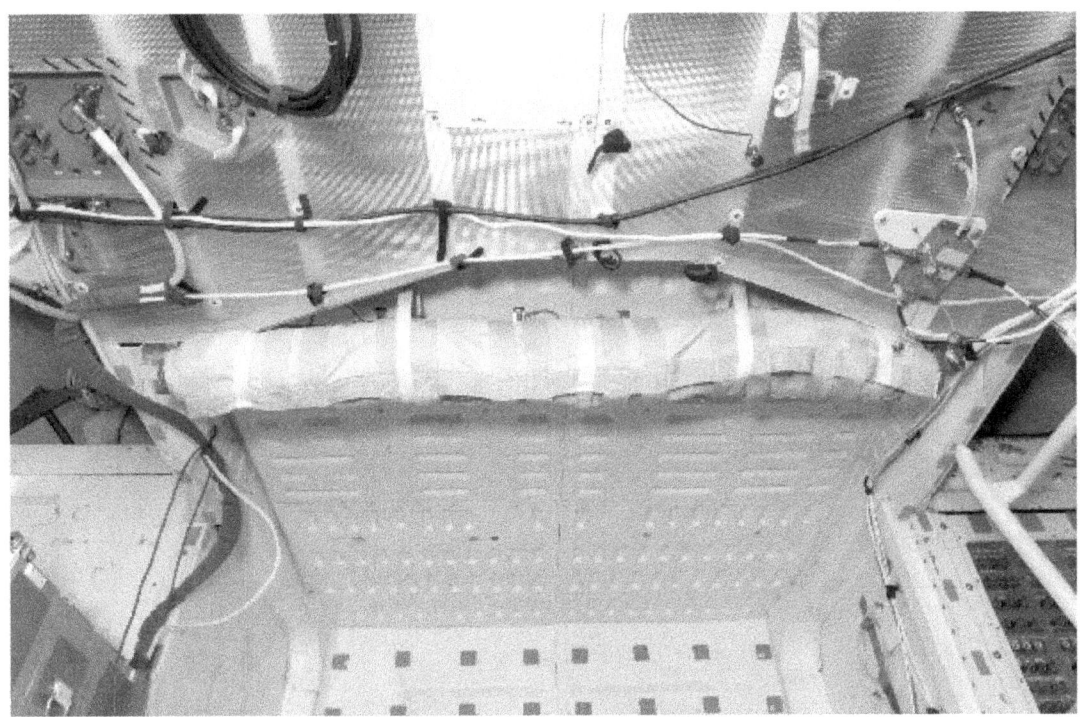

Figure 3-14. Trampoline – stowed

During postinsertion, the crew will perform the following procedures to stow the trampoline:

Step	Action
1	Release the port, ceiling, and deck closeout net and disconnect the starboard trampoline Nomex extension.
2	Unlock the ratchet mechanism at the ceiling by rotating both release knobs in the clockwise direction, which will loosen the trampoline fabric.
3	Disengage the three stowage straps from the three D-rings located on the top of the trampoline assembly.
4	Remove the three straps from the interdeck access ladder via the "pull-the-dot" snaps.
5	Pull the six double-acting pins from the floor fittings.
6	Roll the trampoline from the floor to the ceiling with the webbing on the inside. Roll the trampoline with the closeout nets inside the assembly.
7	Wrap the stowage straps around the trampoline and retain with D-rings.
8	Unfurl the two retention straps and secure trampoline to ceiling with retention straps.

3.4 SIDE HATCH

3.4.1 Introduction

The side hatch, shown in Figure 3-15 from the interior, is the primary means of entering and exiting the orbiter.

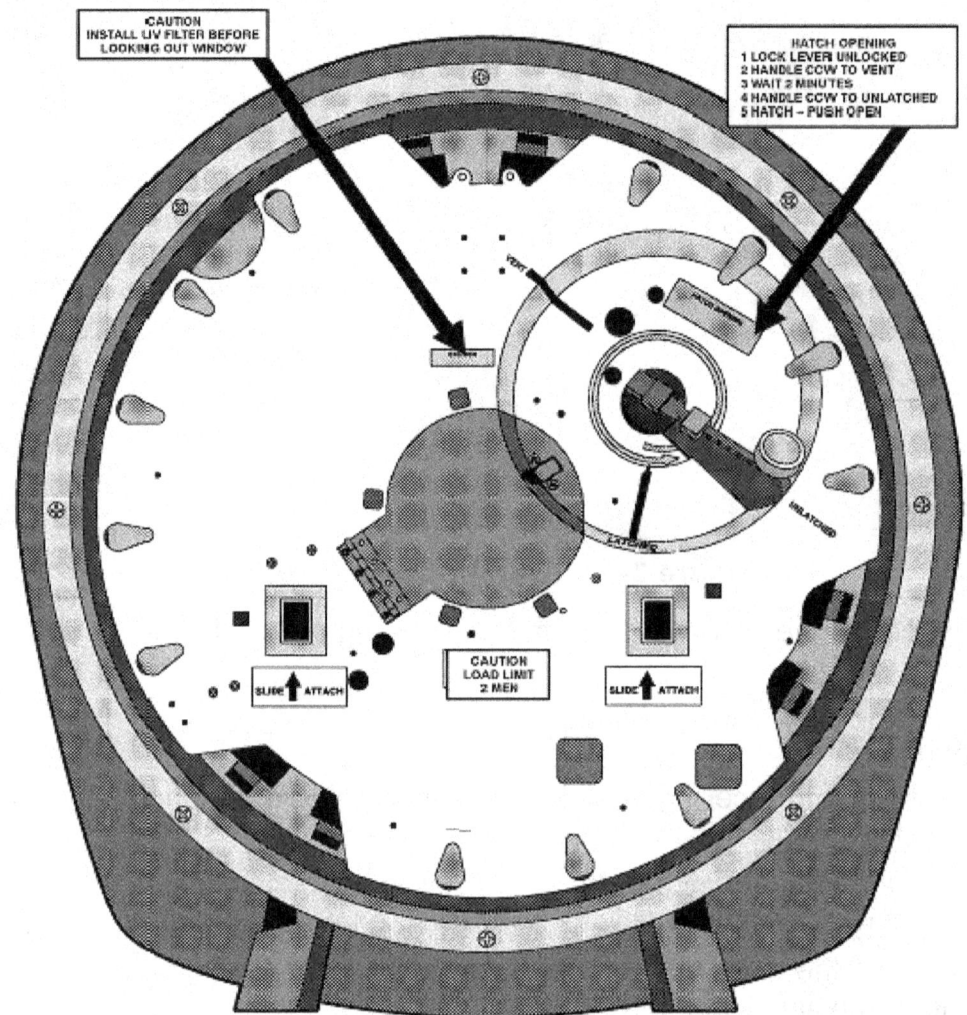

Figure 3-15. Orbiter side hatch, interior view

3.4.2 Physical Features

The physical features of the side hatch are as follows:

a. Weighs ~300 lb.

b. Has 40-in. diameter opening.

c. Has a 10-inch diameter optically pure window in the center.

d. Is attached to crew cabin structure by hinges and secured closed by 18 hatches.

e. Has two pressure seals.

3.4.3 Functional Features

The functional features of side hatch mechanisms are as follows:

a. Force required to operate the hatch mechanisms:

   1. Lock lever: 6 to 13 lbf to unlock.

   2. Hatch handle: ~16 lbf to rotate, 25 lbf to overcome vent position detent.

b. A hydraulic attenuator controls the rate of hatch opening.

c. Hatch opens outwardly 90° (down when vehicle is in horizontal position).

3.4.4 Normal Side Hatch Opening

During normal operations, the flight crew does not need to operate the side hatch because:

a. Prelaunch, the closeout crew closes and seals the hatch.

b. Postflight, the convoy crew opens the hatch.

However, because crewmembers may need to open the side hatch for a prelaunch or postlanding emergency, they need to be familiar with the nominal hatch opening procedure. The procedure is supplied on a decal affixed to the side hatch (Figure 3-17) and is iterated below.

Step	Action
1	Ensure that the side hatch-locking device is removed.
2	Flip lock lever on hatch handle to UNLOCKED position.
3	Rotate hatch handle counterclockwise to VENT position. **WARNING** Stop at VENT position for 2 minutes to avoid rapid hatch opening.
4	Rotate hatch handle counterclockwise to UNLATCHED position (hard stop).
5	Push hatch open.

3.4.5 Depress, Hatch Jettison

Both the depressurization and hatch jettison are accomplished through a pyrotechnic system that includes the following components:

3.4.5.1 T-Handle Pyro Box Feature/Function

a. Located on port forward middeck floor.

b. Equipped with safing pin.

c. Has two T-shaped handles:

   1. <u>Vent valve T-handle</u>:

      (a) Safing pip pin (remove before vent)

      (b) Is aft, smooth handle (squeeze and pull, requires 20-50 lbs to release).

      (c) Has no cloth cover.

      (d) Activates pyro vent valve to depressurize cabin (in-flight bailout only).

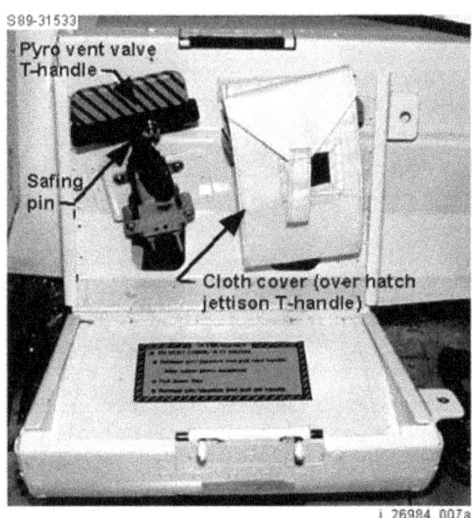

WARNING

Initiating cabin vent on ground could be an ignition source in payload bay.

2. Hatch jettison T-handle:

   (a) Safing pip pin (remove before jettison)

   (b) Is forward, ribbed handle (squeeze and pull, requires 20-50 lbs to release).

   (c) Has white cloth cover.

   (d) Activates hatch pyrotechnics to jettison hatch.

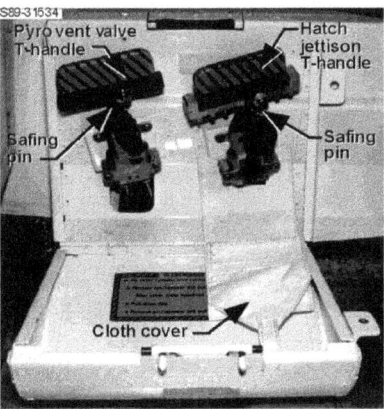

CAUTION

During in-flight bailout only, cabin pressure must be equalized before hatch is jettisoned. Otherwise, middeck structural damage will likely result.

Each T-handle has a safing pin (see illustration above) that must be removed before handle can be moved and squeezed.

The handles are connected to respective pyro components by redundant lengths of shielded mild detonating cord (that require no orbiter power).

3.4.5.2   Pyro Vent Valve Feature/Function

Located behind waste management compartment (port side of middeck aft bulkhead):

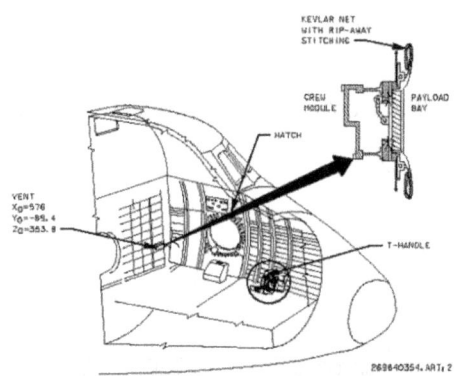

The vent valve depresses the cabin at a controlled rate by using redundant linear shaped charges to blow out a plug allowing air to pass through two holes with a combined area of ~15 in^2 between the crew compartment and the payload bay.

> **WARNING**
>
> The cabin vent pyros are an ignition source in the payload bay and should not be used for postlanding egress.

3.4.5.3 Hatch Pyrotechnics Feature/Function

The pyrotechnics associated with the hatch include:

a. Four linear-shaped charges (two per hinge for redundancy) to sever the hinges.

b. Two (redundant) expanding tube assemblies to sever the 70 frangible bolts holding the hatch adapter ring to the orbiter.

c. Three (partially redundant) thruster packs to separate the hatch from the orbiter at a velocity of approx. 45 ft/sec (30 mph).

Note: The hatch jettison features could be used in a bailout or landing emergency.

3.4.6 Side Hatch Opening by Rescue Personnel

In an emergency prelaunch or postlanding, rescue personnel can open the side hatch from the outside by using the external side hatch opening/closing tool shown in Figure 3-16 (or by using a 1/2-in. or 5/8-in. square drive with a 12-in. extension).

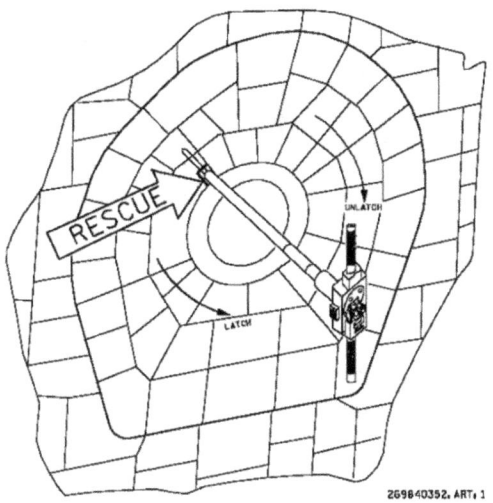

Figure 3-16. External side hatch opening/closing device

Rescue personnel follow the procedures below, presented here for general awareness, to open the orbiter side hatch from the outside:

Step	Procedure	
1	**WARNING** Following a nominal entry, hatch window temps can exceed 200° F 4 min after touchdown and 130° F 30 min after touchdown. On side hatch, punch through the small red tile to expose the latch mechanism receptacle.	
2	Insert the hatch release tool into the latching mechanism receptacle.	
3	Strike the end of tool with enough force to disengage the lock lever on the internal release mechanism.	
4	Rotate the latch release mechanism clockwise in two stages:	
	Stage 1	Rotate to the VENT detent; wait for pressure equalization. (Nominal wait is 2 min. Worst-case wait is 30 sec at sea level and up to 1 min at higher altitudes).
	Stage 2	Rotate to hard stop (UNLATCH position).

Step	Procedure
	WARNING Stand clear when opening the hatch. The side hatch weighs 300 lb. Positive cabin pressure can cause the hatch to initially open rapidly.
5	If necessary, pull or pry the hatch open from the top (hatch opening is attenuated).

3.5   WINDOW 8 ESCAPE PANEL

3.5.1   Introduction

Postlanding, if emergency egress through the side hatch is impossible, the crew uses the secondary route, which is via the Window 8 escape panel (Figure 3-17).

To use this secondary route, crewmembers

a. Jettison the Window 8 escape panel using the panel pyrotechnic system as described in subsection "Escape panel jettison."

b. Attach to a sky genie (see next section).

c. Climb out Window 8 using MS2's seat.

d. Descend the starboard side of the orbiter (Can descend on port side, but starboard side is preferable to avoid side hatch pyrotechnics).

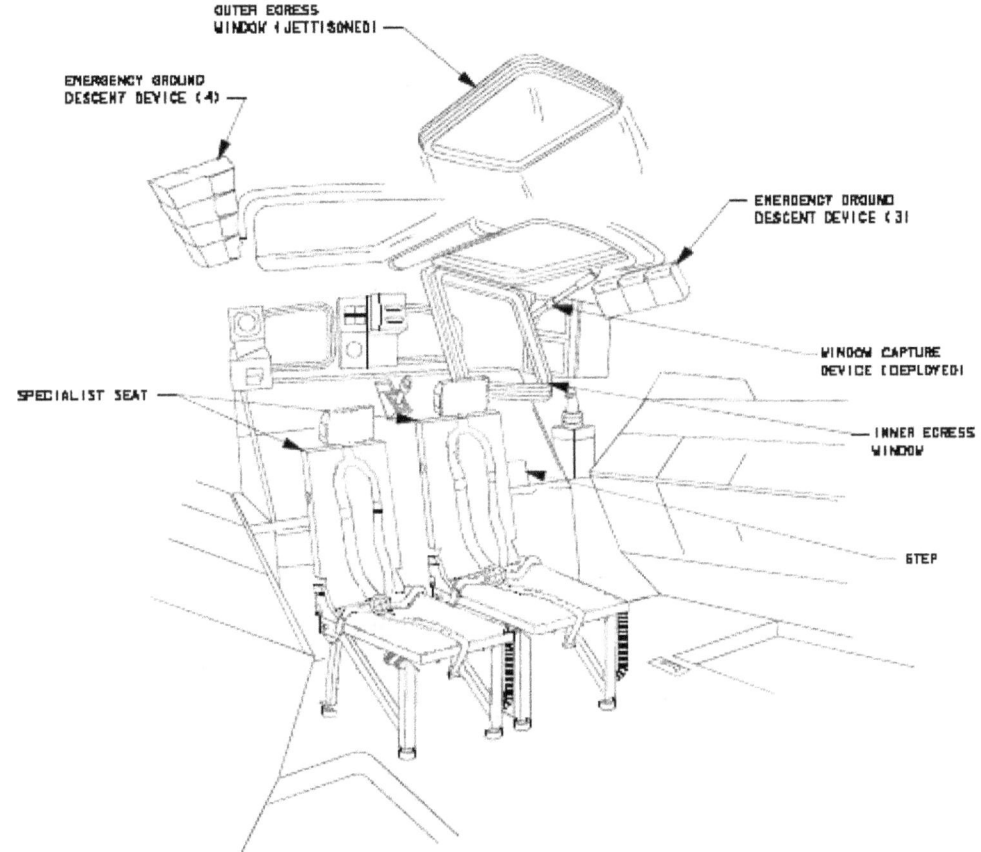

Figure 3-17. Window 8 escape panel

3.5.2 Panel Pyrotechnics

The pyrotechnic system for jettisoning the escape panel is:

a. A pyrotechnic system requiring no orbiter power to operate.

b. Initiated by the jettison ring handle (Figure 3-18) located on top of flight deck panel C2 (center console).

c. Initiated by ground crew via external handle on starboard side of the orbiter forward of the wing.

Pyro components include expanding tube assemblies, mild detonating fuses, frangible bolts, and associated initiators.

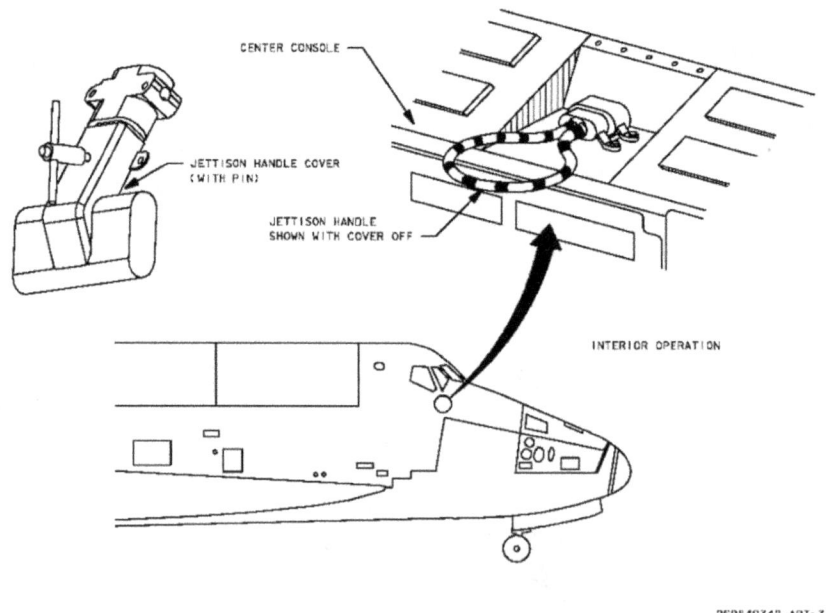

Figure 3-18. Escape panel jettison handle, panel C2

3.5.3 Escape Panel Jettison

The operational detail of opening the escape panel is as follows:

Step	Action
1	As the handle initiator is pulled, the outer thermal windowpane and frame are ejected upward and overboard as a unit. **CAUTION** The thermal pane window shatters. Take care to avoid injury from glass fragments.
2	At 0.3 sec after the outer pane is jettisoned, the two inner pressure panes hinge down and aft and rest on the aft flight deck panel as illustrated below: a. A <u>capture device</u> prevents the inner windowpane from opening with too much force and holds the panel open. b. A <u>no-back device</u> keeps the inner window from closing when the pressure across it is too great (as can happen at a landing field with a high elevation). This no-back device is spring loaded to: 1. Follow the window during the initial opening. 2. Engage a ratchet to prevent subsequent closing.
3	Should the inner window fail to open (due to positive delta pressure), a prybar is used to force the window open (see next subsection).

3.5.4 Manual Opening with Prybar

If the pyro system fails to open the escape panel against delta pressure, the crew can open the panel manually with a prybar (locker A17).

Note: The pyro system must be used prior to using the prybar.

As shown in Figure 3-19 below, there are two pry locations on the escape panel, each location color-coded for the end of the prybar to use:

Primary pry location

Marked green.

Use if panel opens slightly.

Use green end of prybar.

Backup pry location

Marked yellow.

Use if panel does not open at all.

Use yellow end of prybar.

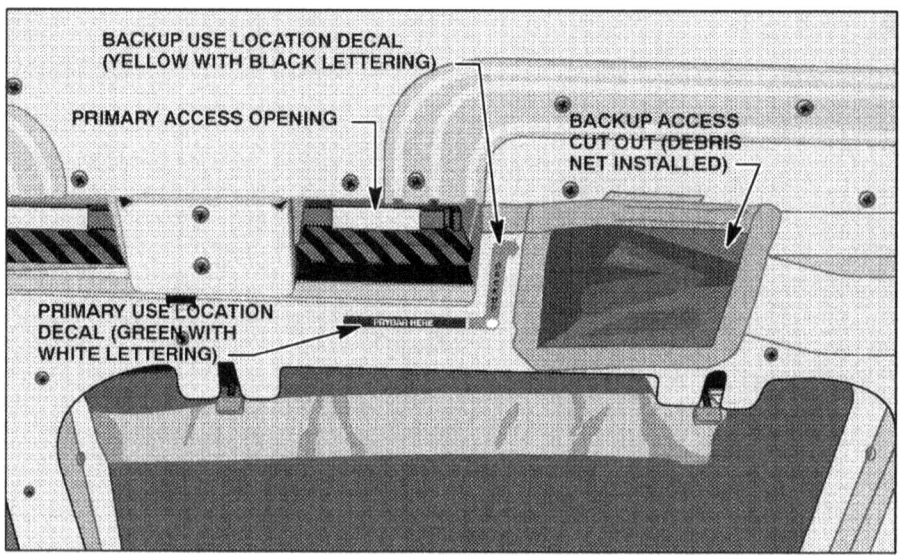

Figure 3-19. Escape panel pry locations

Figure 3-20 below shows prybar positioning in the primary and backup locations:

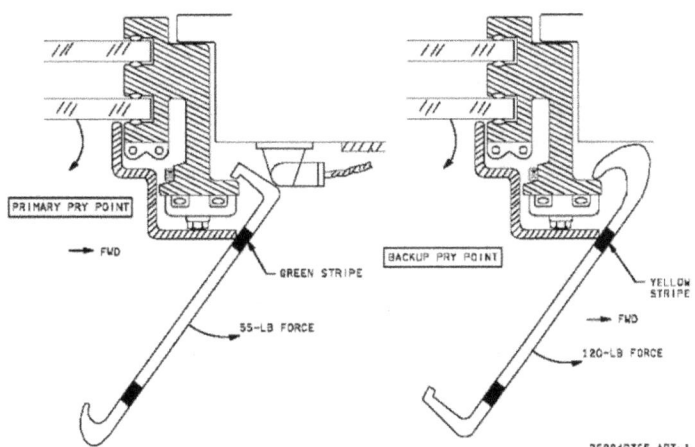

Figure 3-20. Prybar positioning

3.5.5 Opening by Rescue Personnel

In a postlanding emergency when entry through the side hatch is impossible, rescue personnel can open the Window 8 escape panel from the outside to enter the orbiter and aid the orbiter crew. The external T-handle initiator, located on the starboard side of the orbiter forward of the wing, is depicted in Figure 3-21.

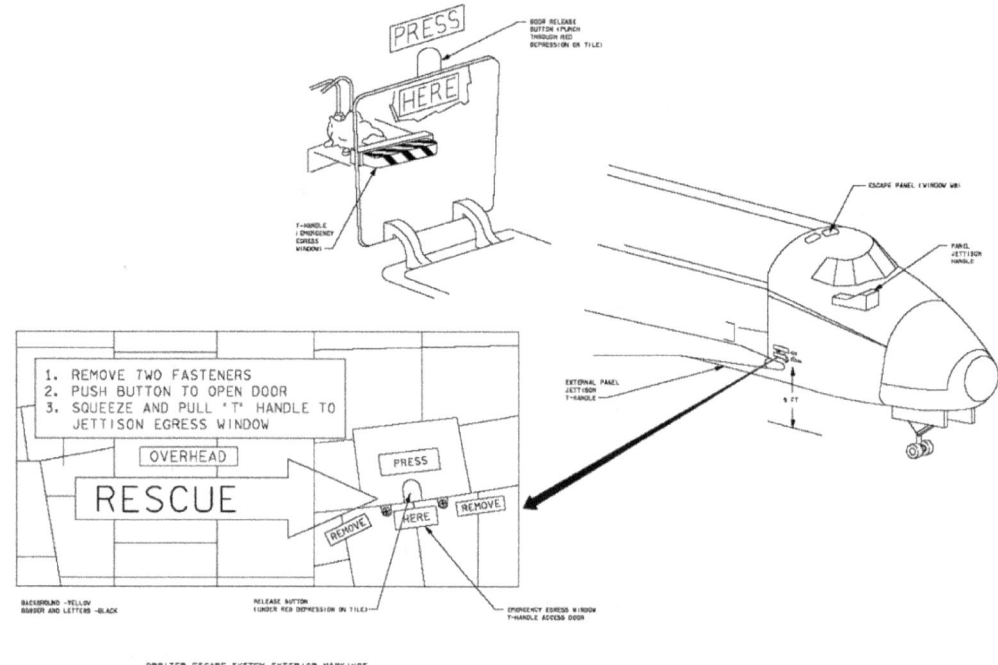

Figure 3-21. External T-handle initiator

Rescue personnel follow the procedures below, given here for general awareness, to open the escape panel from the outside to rescue the orbiter crew.

Step	Procedure
1	**WARNING** a. Advise flight crew and crash/rescue personnel of intent to jettison the emergency escape panel and to stay clear of inner egress window opening path (ground comm). b. Before jettisoning the escape panel, fire-fighting personnel should direct a water fog across the top of the orbiter to dissipate any explosive/flammable atmosphere that may have accumulated in the panel area. On the lower right side of the forward fuselage: a. Remove two fasteners. b. Punch through small red tile. c. Depress door release button and open.

Step	Procedure
2	Squeeze and pull Window 8 jettison T-handle (handle pulls free of initiator assembly when squeezed and pulled).
3	Using a ladder or other suitable means, gain access to top of orbiter. **WARNING** Following a nominal entry, the thermal protection system in the overhead window area can be 120° F 4 min after touchdown, and the overhead window glass can be 190° F.

If outer window...	Then...
a. Broke loose from the orbiter but failed to jettison overboard	Pry window loose at its forward edge with a prybar; lift window and toss overboard.
b. Has pyro failure	Break thermal pane with prybar or heavy hammer.
If inner window...	**Then...**
a. Closed after jettison operation	• First, apply pressure with prybar at forward end of window to open and allow internal/external pressures to equalize. • Next, manually push window down to fully open position.
b. Has pyro failure	Proceed to break both panes of the window.

Step	Procedure
4	Enter aft flight deck through window opening.
5	Immediately verify that all crewmembers have: a. Released their neck dam tabs. b. Closed visors. c. Activated emergency O_2.
6	Assess conditions (fire, debris) in cabin. If fire or smoke evident, activate fire suppression system and/or use portable fire extinguishers as required. Note: Orbiter power must be available to activate the fire suppression system.
7	Power down orbiter electrical systems.
8	Aid or rescue each crewmember, beginning with MS directly beneath Window 8: a. Crewmembers requiring little or no assistance use descent devices. b. Incapacitated crewmembers will be lifted up through window opening and lowered to ground by rescue personnel by means available.

3.5.6 Cut-in Area

If in extreme circumstances it is not possible or advisable to open either the side hatch or the Window 8 escape panel, rescue personnel can remove a section of the orbiter wall (Figure 3-22) to gain access to the middeck. Removal of the section, called the "cut-in" (or "cut-out"), requires 45 min or more. (Figure 3-22 also shows the secondary interdeck access.)

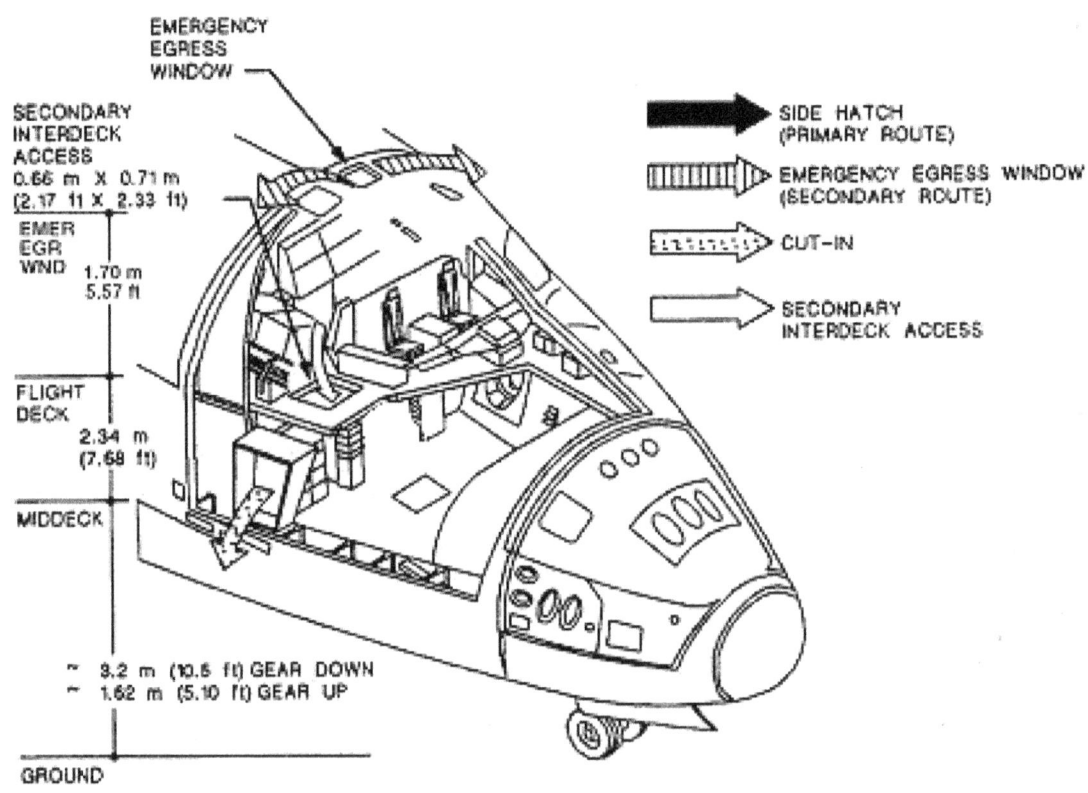

Figure 3-22. Cut-in area

Note: Overall, from top to ground is 7.24 meters, or 23.75 ft.

3.6 DESCENT CONTROL DEVICE (SKY GENIE)

3.6.1 Introduction

Crewmembers can use the descent control device, or "sky genie," in an emergency postlanding situation to lower themselves from the orbiter to the ground quickly and safely in either of two ways:

1. <u>If the egress slide fails</u>, through the side hatch (either opened normally or jettisoned).

2. <u>If the side hatch fails altogether</u>, through the Window 8 escape panel.

3.6.2 Functional Features

Each sky genie, depicted in Figure 3-23, includes the following components:

a. A descent device

b. 40 ft of nylon rope

c. Rope bag

d. Crewmember tether and hand loop

e. Swivel snap shackle

f. Emergency release tab

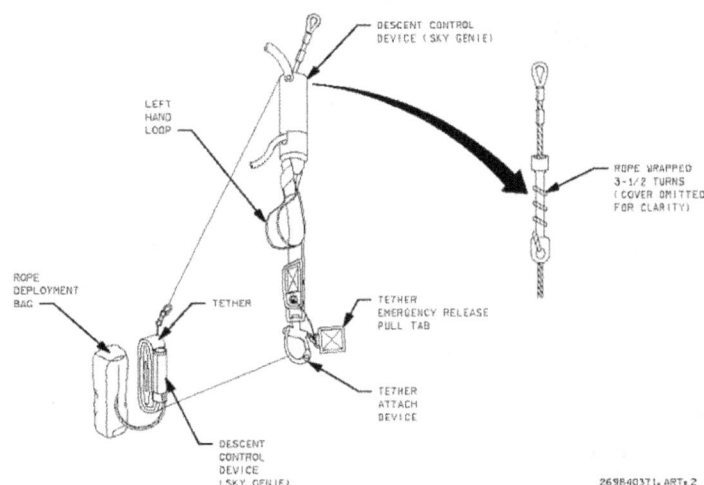

Figure 3-23. Sky genie components

3.6.3 Sky Genie Use

Sky genie use is depicted in Figure 3-24.

Figure 3-24. Sky genie use

The nylon rope is attached to a support bracket forward of the escape panel by a metal cable. The rope is fed through an opening at the top of the descent device and wrapped 3 1/2 times around the central shaft (the friction of the rope passing around the shaft allows the crewmember to control the descent rate). Refer to Figure 3-23. The rope exits the device on the same side as the crewmember's tether and hand loop. (A metal cover over this end of the device keeps debris out and provides a smooth sliding surface.) The free length of the rope is stowed in a bag. To use, crewmembers throw the bag containing the free length of rope out the egress opening. The rope's weight helps unfurl the rope when the bag is thrown. Crewmembers attach sky genie snap shackle to the carabiner on their harness. At this point, the crewmembers can let go of the sky genie and use two hands to egress.

3.6.4 Sky Genie Use, Side Hatch

Since the side hatch is the primary egress route from the orbiter, the crew uses the sky genie through the side hatch in case of egress slide failure. The procedure is as follows for using the sky genie through the side hatch:

Step	Action
1	Jettison or open side hatch manually.
2	A crewmember on the flight deck passes down a sky genie rope.
3	The rope is thrown out the hatch opening (aft of the hatch).
4	A sky genie must be manually pulled down from the flight deck and attached to the crewmember's carabiner.
5	The crewmember crawls out the side hatch opening, places his feet on the side of the orbiter, and lowers himself to the ground.

3.6.5 Sky Genie Use, Window 8 Escape Panel

The procedure for using the sky genie in egressing through the Window 8 escape panel is as follows:

Step	Procedure
1	Pull Window 8 jettison handle.
2	Remove sky genie from stowage bag and free cable from cable track. **CAUTION** Take care not to drop the rope in the cabin. Dropping it could put knots in the rope or cause shattered glass and debris to stick to it, making descent difficult or impossible.
3	Throw sky genie rope out Window 8 over **starboard** side of orbiter (descend vehicle on starboard side because hatch could still jettison unexpectedly).
4	Uncover and open harness carabiner and attach sky genie snap shackle to carabiner.
5	**CAUTION** Exercise caution in egressing the Window 8 opening to avoid entanglement and injury from metal cables and ropes. **WARNING** Following a nominal entry, the thermal protection system in the overhead window area can be 120° F 4 min. after touchdown, and the overhead window glass can be 190° F. Check to ensure egress step is deployed, then climb onto MS2's seat and through Window 8. Note: Make sure that sky genie cable is **over left shoulder** when climbing through the window area. Positioning the cable thus keeps cable from wrapping around you.
6	While sitting on windowsill, check that the rope and cable are not tangled or caught on any part of suit or around arm or leg.

Step	Procedure
7	Hold onto the sky genie hand loop with the left hand and onto the loose end of the rope with the right hand.
8	While lying on your left side, slowly slide down the starboard side of the orbiter until the sky genie cable and rope are taut. Note: The starboard side is recommended, thus staying clear of potential hatch jettison on the port side. Of course, if a fire or hazard is present on the starboard side, the port side must be used.
9	Proceed to lower self down the starboard side of the orbiter while controlling the rate of descent by applying tension on the sky genie rope.
10	Upon reaching the ground, disconnect the snap shackle from the carabiner and assist the next person in descending.
11	Proceed upwind of orbiter to avoid contaminants and to remain visible to support crew.

3.7 ESCAPE POLE

3.7.1 Introduction

The crew escape system orbiter-based hardware includes the escape pole, depicted in Figure 3-25. The pole is designed to guide crewmembers during bailout on a trajectory that clears the orbiter's left wing.

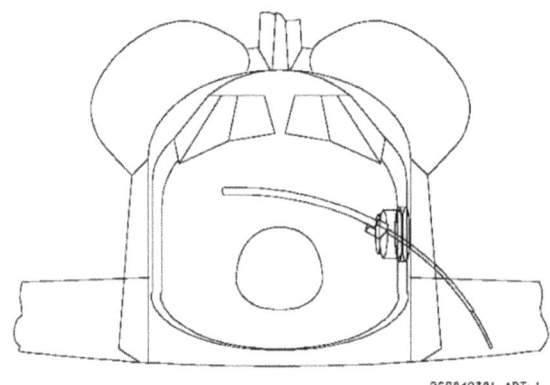

Figure 3-25. Deployed escape pole (looking aft)

3.7.2 Functional Features

The functional features of the escape pole are as follows:

a. Spring-activated, telescoping curved steel pole.

b. Extends outward from side hatch 9.2 feet when fully deployed.

c. Is contained within an aluminum housing that attaches to the middeck ceiling and the port wall forward of the hatch.

d. Can be removed from the ceiling and stowed on orbit.

e. Has a magazine holding eight lanyards attached to port (hatch) end. The lanyards guide the crewmembers down the pole.

f. Weighs 267 lb (the entire assembly, including pole and lanyards).

3.7.3 Normal Escape Pole Deployment

The instructions for normal pole deployment are supplied on a decal next to the pole deploy handle, shown in Figure 3-26, and described below.

Step	Action
1	Remove safing pin by pulling upward.
2	Remove arming pin.
3	Rotate deployment handle counterclockwise to stop. Check to ensure green stripe is visible on deployed pole.

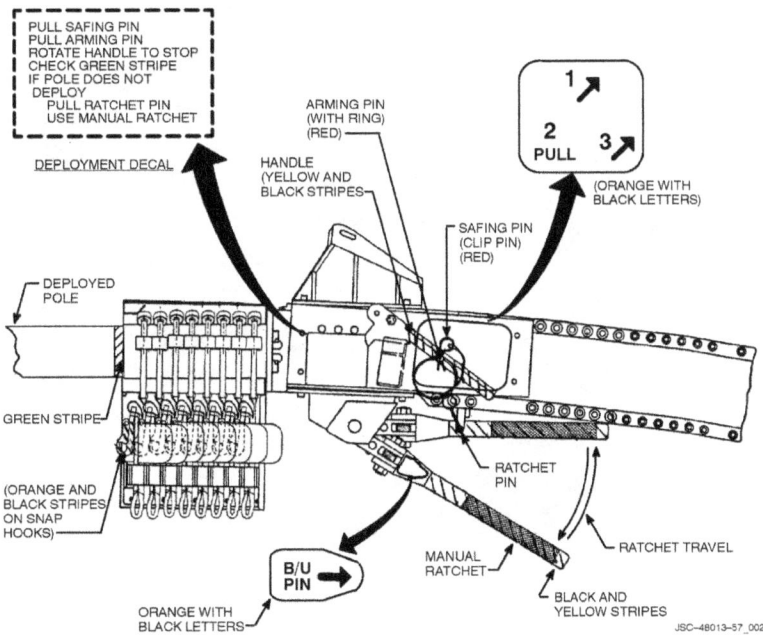

Figure 3-26. Escape pole deployment

3.7.4 Manual Escape Pole Deployment

Should the pole fail to extend fully in normal deployment (green stripe not visible, see Figure 3-26), the ratcheting device shown in Figure 3-26 is used to extend the pole manually.

Step	Action
1	Remove ratchet pin from ratchet handle.
2	Cycle handle up and down to extend pole.
	Each upstroke of the ratchet handle extends the pole ~4 in., taking ~25 strokes to deploy fully. Continue to use ratchet handle until green stripe is visible.

3.7.5 Magazine and Lanyards

The magazine containing a total eight lanyard assemblies (regardless of crew size) is shown in Figure 3-27 below:

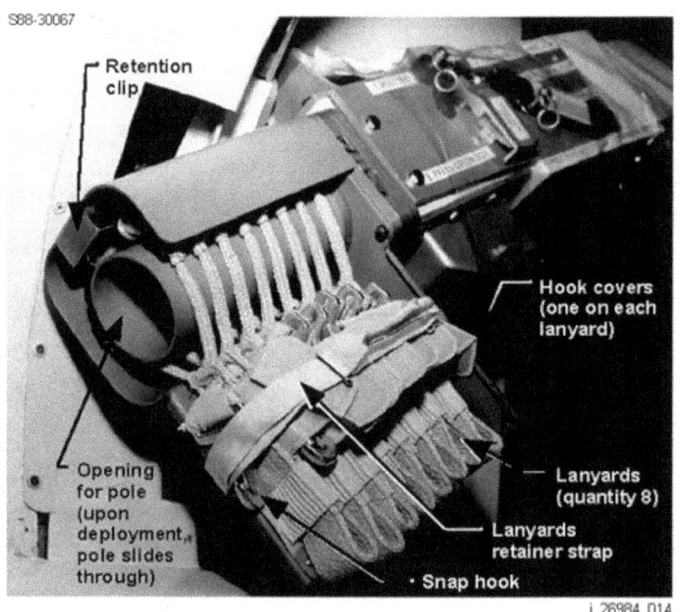

Figure 3-27. Magazine and lanyards

Features to be noted (Figure 3-27):

a. The lanyard retainer strap is held in place with Velcro and is easily removed.

b. The small fabric tab on each lanyard (called a hook cover) lessens the possibility of another crewmember from attaching to any of the remaining lanyards except the outermost one in the magazine.

c. A spring-loaded retention clip at the top of the magazine and an elastic strap at the bottom retain the lanyards.

d. Each lanyard assembly consists of a hook attached to a Kevlar strap that has four roller bearings surrounding the pole.

3.7.6 Lanyard Use

Once the pole is deployed, either normally or manually, lanyard use is as follows:

Step	Action
1	Each crewmember should perform the following before leaving seat: a. Uncovers the lanyard attach ring on right riser of the parachute. b. Pulls the ring out to extend to the attach bridle.
2	The first crewmember to the side hatch removes the lanyard retainer strap by pulling on the cloth loop.
3	Each crewmember: a. Leans forward into the hatch area. b. Attaches the metal ring to the snap hook on the outermost lanyard before bailout.

3.8 EMERGENCY EGRESS SLIDE SYSTEM

3.8.1 Introduction

The Emergency Egress Slide System (EESS) provides rapid and safe orbiter egress during postlanding contingency or emergency situations. The slide system accommodates:

a. Egress to the ground within 60 sec after the side hatch is fully opened (or jettisoned).

b. Removal of incapacitated crewmembers.

Figure 3-28 shows the slide deployed.

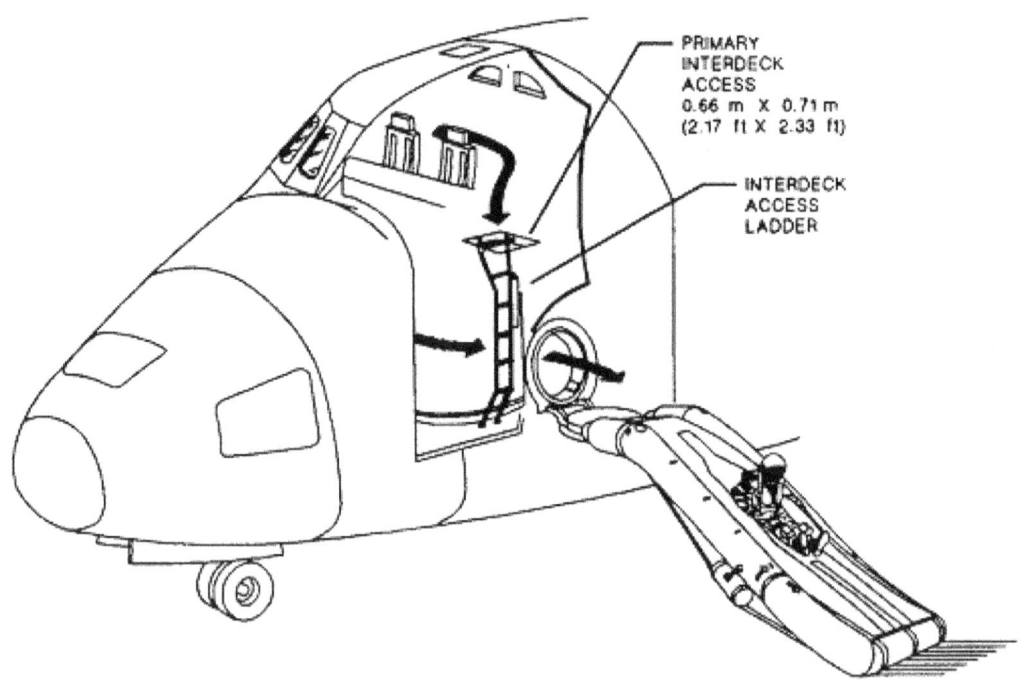

Figure 3-28. Emergency egress slide deployed
(side hatch opened normally)

3.8.2 Physical Characteristics

The slide system consists of the following components:

3.8.2.1 Inflatable Slide

Constructed of neoprene-coated nylon fabric with a sliding surface of urethane-coated nylon fabric.

Inflated by an argon bottle (see next item) to ~2.75 psig. A vent valve, with adjustment screw, bleeds off pressures greater than 3.0 psig.

Remains functional for at least 6 min after inflation.

3.8.2.2 Pressurized Argon Bottle

Pressurized to 3000 psig; equipped with pressure gauge.

When actuated by pulling the inflation lanyard, argon flows at 400 psi through a hose to an aspirator. The aspirator pulls in ambient air (approximately four times the volume of the argon) to inflate the slide with an air and argon mixture in ~3 sec.

3.8.2.3 Slide Support Assembly

Consists of two aluminum plates and brackets attached to wall below side hatch.

Inflatable part of the slide is attached to the support assembly by fabric that is attached to the support assembly. The inflatable portion can be detached from the support by pulling on detach lanyard.

Support assembly (with slide attached) is easily removed and attached to hatch.

3.8.2.4 Slide Container

Made of aluminum; equipped with handle for easy removal.

Attached to orbiter floor and port wall below side hatch by four pins.

Equipped with window for viewing argon bottle pressure gauge.

3.8.2.5 Chem Lights

Two high intensity lightsticks are located above the side hatch.

3.8.2.6 Reingress Strap

The crew module can be reingressed by climbing up the slide using the reingress strap, Velcroed on the aft side of the slide.

3.8.3 Slide Deployment

The egress slide can be deployed in two ways:

a. With the hatch opened normally.

b. With the hatch jettisoned.

3.8.4 Slide Deploy, Hatch On

Slide deployment with the hatch on (hatch opened normally) is as follows:

Step	Procedure
1	Rotate hatch handle counterclockwise to VENT position.
2	Pull pin to release cover handle (plastic tie wraps hold pin in place).
3	Lift handle and remove slide cover (pass cover back to crewmember behind, or toss out of the way).

Step	Procedure
4	Rotate slide 270° into hatch tunnel.
5	Remove the two hinge pip pins; lift entire slide assembly and rotate 90° to insert brackets into the slide attach points on the hatch.
6	Rotate hatch handle counterclockwise to the UNLATCHED position and push hatch open.

USA009026
Basic

Step	Procedure
7	Push the slide off the outboard edge of the hatch. **CAUTION** Side hatch is approximately 10 ft above the ground.
8	Pull the inflation lanyard all the way out **without releasing grip**. Note: Premature release of inflation lanyard may result in slide not inflating.

3-38

Step	Procedure
9	Crawl onto the side hatch; descend slide feet first. Before descending: • Have legs and feet slightly apart. • Put hands at side of body to aid balance. • Lean back with weight off heels. • Prepare to stand when reaching bottom. There is a two-person load limit on the side hatch (includes person on slide).

USA009026
Basic

3.8.5 Slide Deploy, Hatch Off

Slide deployment with the hatch off (hatch jettisoned) is as follows:

Step	Action
1	Pull pin to release cover handle (plastic tie wraps hold pin in place).
2	Lift handle and remove slide cover (pass cover back to crewmember behind, or toss out of the way).
3	Rotate slide 270° into hatch tunnel.
4	Push the slide pack overboard.

3-40

Step	Action
5	Pull the inflation lanyard all the way out **without releasing grip**. Note: Premature release of inflation lanyard may result in slide not inflating.
6	Grab the handhold above the hatch tunnel; swing legs out and descend the slide feet first. Before descending: - Have legs and feet slightly apart. - Put hands at side of body to aid balance. - Lean back with weight off heels. - Prepare to stand when reaching bottom. There is a two-person load limit on the side hatch (includes person on slide).

4.0 NOMINAL CREW SEAT PROCEDURES

4.1 OVERVIEW

4.1.1 Introduction

There are numerous procedures that must be completed during nominal seat ingress/egress. Seat ingress/egress will occur during:

a. Prelaunch

b. Postinsertion

c. Deorbit Prep

d. Postlanding

4.1.2 Purpose

This chapter describes steps and procedures required for nominal seat ingress/egress.

4.1.3 Objectives

After completing this chapter, you will be able to identify:

a. Steps for prelaunch seat ingress.

b. Steps for postinsertion seat egress.

c. Steps for deorbit prep seat ingress.

d. Steps for postlanding seat egress.

In This Chapter

Section describes seat ingress/egress as follows:

Section	Page
Nominal Seat Ingress/Egress	4-1

4.2 NOMINAL SEAT INGRESS/EGRESS

Nominal seat ingress and egress procedures are defined by the mission phase during which the ingress or the egress occurs.

Ingress and egress during each mission phase are discussed below. For seat/orbiter egress because of a mission scrub, see "Nominal seat/orbiter egress" at the end of the section.

4.2.1 Prelaunch Seat Ingress

The sequence of events is as follows immediately before and during prelaunch ingress (T – 3:45:00 to T – 1:30:00):

Pre-ingress:

Time	Step	Action
~ T - 3:45	1	Don flight suits and conducting pressure checks in the Operations and Checkout (O&C) building suit room.
~ T - 3:15	2	Crew Transport to 195-foot level of the Fixed Service Structure (FSS).
~ T - 2:45	3	In the white room on the Orbiter Access Arm (OAA), the Closeout Crew and the suit technician assist the crew in donning a parachute harness and a CCA (comm cap).

Ingress:

Time	Step	Action
~ T - 2:25	1	Crewmembers ingress through the orbiter side hatch into the flight deck and middeck to assigned seats in the following order (ingress into the two areas can be at the same time): **Flight Deck** **Middeck** CDR (Seat 1) MS5/PS2 (Seat 7) PLT (Seat 2) MS3 (Seat 5) MS1 (Seat 3) MS4/PS1 (Seat 6) MS2 (Seat 4)

Time	Step	Action
~ T - 2:25	2	A suit technician (insertion technician), aided by Astronaut Support Personnel (ASP), assists each crewmember into assigned seat. • The seat has a parachute pack on it (pack acts as seat back cushion). • Crewmembers lie on their back on the parachute.

	3	The suit technician and the ASP help the crewmembers with strap-in (flight deck crew and middeck crew at same time).
		
		The Commander (CDR) and the Pilot (PLT) have additional procedures for adjusting the headrest, rudder pedal, and Rotational Hand Controller (RHC); otherwise, the strap-in procedures are basically the same for all crewmembers.
		
		These strap-in procedures are outlined below:
		a. Normally, parachute connections are made first, and in the following order:
		   1. Two upper Frost fittings
		   2. Lower ejector snaps
		b. The cooling may be attached and turned on to cool the crewmember.
		c. The shoulder harness/lapbelt assembly is connected in the following sequence:
		   1. Lapbelt is connected to the buckle on the crotch strap
		   2. Shoulder harness is attached
		
		When connections are completed, the shoulder harness and lapbelt are adjusted.
		d. The helmet is placed on the neck ring and securely fastened.
		e. The orbiter O_2 hose is attached.
		f. The helmet comm leads and the HIU are connected.
		g. Don gloves.

Time	Step	Action

~ T - 2:25 (continued)	3 (continued)	The ASP conducts suit pressurization and comm tests (flight deck crew first, middeck crew second):
		
		a. The ASP asks each crewmember to perform a suit pressurization test consisting of the following steps:
		   1. Close and lock helmet visor.
		   2. Turn suit O_2 ON.
		   3. Push the push-to-test button on suit dual suit controller.
		
		The test checks the operation of the secondary pressurization system of the dual suit controller and confirms the suit pressurization function.
		
		b. Crewmember communications are checked both with the helmet visor up and with the visor down.

~ T - 1:30	4	Once the strap-in and tests are completed, the ASP and suit technicians leave the vehicle and the orbiter hatch is closed.

4.2.2 Postinsertion Seat Egress

Postinsertion procedures for seat egress follow the activities outlined in the Postinsertion Checklist.

Generally, the procedural sequence is as follows:

Time	Step	Action
~ L + 55 min	1	The mission specialists egress seats first (to begin configuring the aft flight deck and the middeck for orbital operations).
~ L + 1:35 (go for orbit ops)	2	MS3 installs the side hatch-locking device and the pyro box safing pin.
	3	PLT and CDR egress seats.
	4	All crewmembers doff and stow suits and stow the seats.

4.2.3 Deorbit Prep Seat Ingress

In preparing for deorbit, crewmembers install seats and suit up and strap themselves into their assigned seats.

The strap-in procedures are the same as those for prelaunch (see "Prelaunch seat ingress"), except that the crewmembers help each other in ensuring proper configuration and checking that all connections are made.

The procedural sequence is as follows:

Step	Procedure
1	Unstow and install MS seats. Install cooling units on side of seat.
2	If bioinstrumentation manifested, connect leads, cuffs, etc., as necessary, before donning suit.
3	Place parachute on seat.
4	Don suit and parachute harness.
5	Sit in assigned seat and connect parachute (placed on the seat back in step 3).
6	Connect shoulder harness/lapbelt.
7	Connect cooling.
8	Don comm cap and helmet.
9	Connect comm leads and orbiter O_2 hose.
10	Don gloves, kneeboard, etc.
11	If bioinstrumentation manifested, make additional connections if necessary.

4.2.4 Nominal Seat/Orbiter Egress

Nominal seat/orbiter egress occurs in two instances:

a. At the planned end of the mission.

b. Prelaunch, if the mission is scrubbed.

Postlanding

Nominal egress at the planned end of the mission is as follows:

Step	Action
1	At wheel stop, crewmembers unstrap from seats and prepare to egress the orbiter.
2	Convoy crew moves into position, opens side hatch, and assists crewmembers in doffing suit and/or egressing the orbiter.

Prelaunch

In the event that the mission is scrubbed, nominal egress is similar to that for the planned end of the mission. The difference is that prelaunch, the flight crew is assisted by the closeout crew, the ASP, and the suit technicians.

USA009026
Basic

5.0 EMERGENCY EGRESS AND ESCAPE

5.1 OVERVIEW

5.1.1 Introduction

There are eight crew emergency egress/escape modes. These modes are differentiated by:

a. The mission phase during which the emergency situation arises.

b. The nature of the emergency situation.

c. Whether support personnel assist the orbiter crew.

5.1.2 Purpose

This chapter describes the different emergency crew egress/escape modes, the common guidelines, and the orbiter crew response in each mode.

5.1.3 Objectives

After completing this chapter, you will be able to identify:

a. The eight crew egress/escape modes.

b. The guidelines common to several or all modes.

c. The sequence of crew action in each mode.

d. SAR resources available for bailout.

In This Chapter

Sections describe the emergency egress/escape modes as follows:

Section	Page
Emergency Egress/Escape Modes	5-2
Mode I – Unassisted Pad Egress/Escape	5-5
Modes II, III, IV – Assisted Pad Egress/Escape	5-15
Mode V – Unassisted Postlanding Egress/Escape	5-18
Modes VI, VII – Assisted Postlanding Egress/Escape	5-31
Mode VIII – Bailout	5-32

5.2 EMERGENCY EGRESS/ESCAPE MODES

5.2.1 Introduction

The table below summarizes the eight orbiter crew emergency egress modes by mission phase: four modes occurring prelaunch, three postlanding, and one in flight (bailout).

In Modes I, V, and VIII, indicated by a checkmark (√) in the table, the orbiter crew egresses unassisted by support personnel. These three unassisted modes are practiced in training since the crew is expected to egress unassisted.

In the other five modes (II, III, IV, VI, VII), the crew is assisted by the support personnel indicated in the table below.

Table 5-1. Escape Modes

Mission Phase	Type		Mode	Description
Prelaunch	Pad egress and escape	√	I	Orbiter crew unassisted.
			II	Orbiter crew assisted by closeout crew.
			III	Orbiter crew assisted by fire rescue crew.
			IV	Orbiter crew and closeout crew assisted by fire rescue crew.
Postlanding	Orbiter egress and escape	√	V	Orbiter crew unassisted. Egress may be any of three types: a. Hatch on b. Hatch jettison c. Window 8
			VI	Landing on or near runway. Orbiter crew assisted by pre-positioned convoy crew.
			VII	Landing in a remote area. Orbiter crew assisted by rescue and medical personnel arriving by helicopter.
Ascent or entry (for controlled, gliding flight)	Bailout	√	VIII	Bailout from orbiter in following circumstances: a. During ascent or entry b. Over land or water

5.2.2 Assumptions

The following assumptions apply for emergency egress:

a. Regardless of mode, the orbiter crew is expected to egress <u>unassisted</u> by support personnel as much as physical conditions allow.

b. Crewmembers on the short-duration mission will assist returning deconditioned crewmembers in every way possible.

Note: The deconditioned crewmembers will receive all the training and equipment possible to ensure their safety.

c. Persons authorized to call for emergency crew egress are:

   1. Crew Commander (CDR)

   2. NASA Test Director (NTD)

   3. Flight Director (FD)

   4. Convoy CDR

d. All crewmembers must be able to perform all egress tasks (e.g., hatch jettison, cabin vent, escape pole deploy, and escape slide deploy). Thus, if a crewmember assigned to a task is incapacitated, any other crewmember can perform the task without delaying the escape operation.

e. Persons in charge of egress activity for the four prelaunch egress modes are the following:

Mode	Fixed Service Structure	Bunker	Triage	Evacuation
I	CDR	NTD	Triage doctor	Flight doctor
II	Closeout leader	NTD	Triage doctor	Flight doctor
III, IV	Rescue leader	NTD	Triage doctor	Flight doctor

5.2.3 Emergency Egress Cue Card

The emergency egress cue card (Figure 5-1) has one version for all crewmembers.

Subsequent sections detail the procedures for each of these three unassisted egress modes.

Figure 5-1. Emergency egress cue card

A copy of the emergency egress cue card is accessible to individual crewmembers in the following locations:

Crewmember	Cue card location
CDR, PLT	First page of Ascent Flipbook and Go/No-Go Checklist.
MS1, MS2	On back of CDR and PLT seats.
All middeck MSs/PSs	On forward lockers in front of all MSs/PSs.

5.3 MODE I – UNASSISTED PAD EGRESS/ESCAPE

5.3.1 Introduction

Mode I is an <u>unassisted</u> pad egress and escape to a safe place. A Mode I will occur if the emergency situation is too critical to dispatch fire/rescue personnel to the pad and the closeout crew and the ASP have left the launch platform. After the NTD or the CDR calls for Mode I egress, the orbiter crew must do the following without assistance:

a. Unstrap themselves and egress their seats.

b. Open the side hatch.

c. Proceed to the emergency slidewire baskets (Figure 5-2 and Figure 5-3).

d. Descend in the baskets.

e. Proceed to a protective concrete bunker (if directed, evacuate in an M-113 armored personnel carrier).

Note: The OAA (retracted at T-7:00) can be repositioned in ~30 sec.

5.3.2 Slidewire Basket System

The slidewire basket system uses seven separate slidewires and seven multiperson basket assemblies to facilitate rapid escape from the 195-foot level of the FSS to a landing zone 1168 ft away (Figure 5-2 and Figure 5-3).

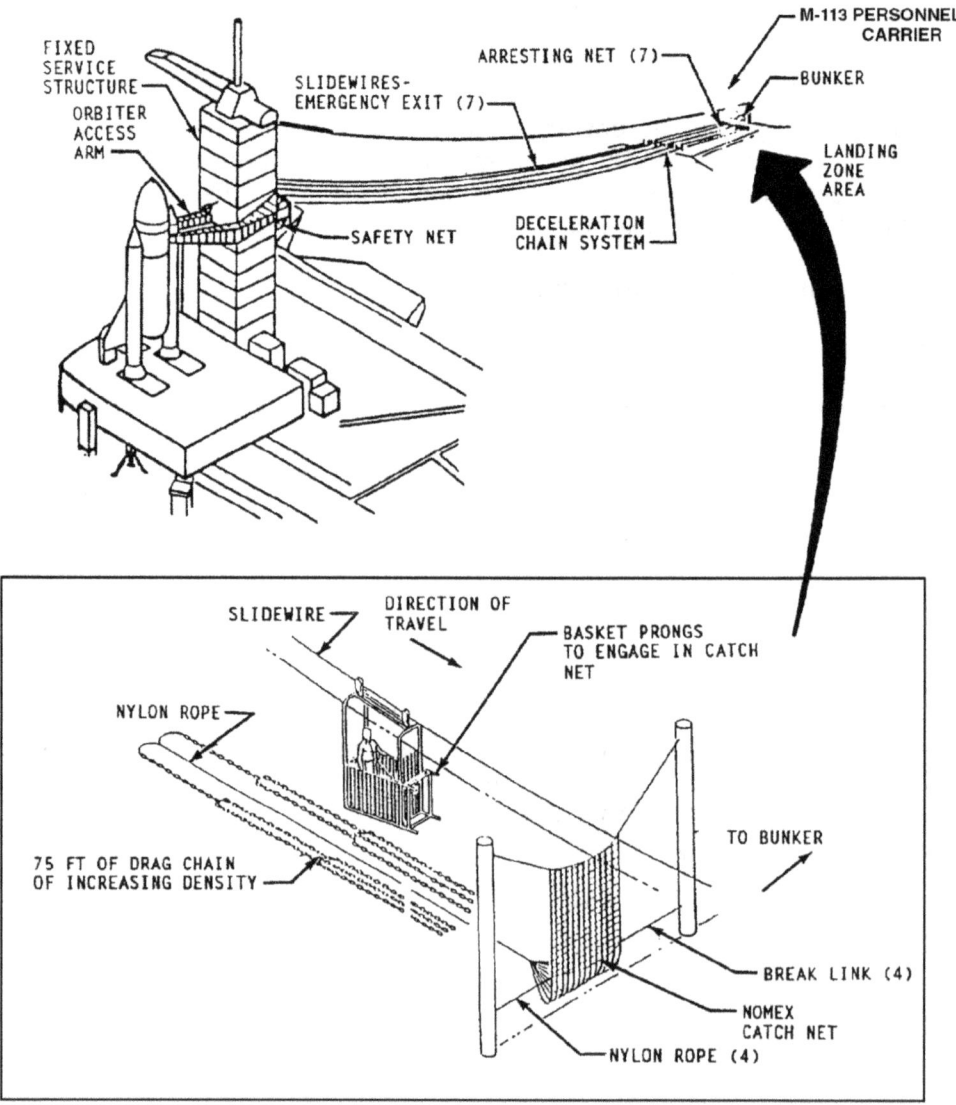

Figure 5-2. Overall view of slidewire basket system

The seven slidewire baskets have the following characteristics:

a. Each can accommodate four persons (with suits on).

b. Hang in recessed cutouts on the west side of the FSS.

c. Are secured in place with Kevlar rope. (A manually operated guillotine-type device is used to sever the rope.)

d. Have fire-resistant material around them to protect passengers from fire below.

e. Are suspended from a 3/4-in. stainless steel cable by two trolleys, one of which is equipped with an antirollback brake.

f. Are slowed by a catch net that is attached to increasingly denser links of chain.

g. The basket south wall releases for exit when a D-ring is pulled.

5.3.3 Sequence of Action

The table below details the sequence of action in Mode I pad egress. The following guidelines apply:

a. All crewmembers follow a buddy system, arranged before flight, and maintain visual or physical contact with the partner at all times (to make sure no one is left behind).

b. If a crewmember has difficulty in performing an assigned task, another crewmember is expected to step in and perform the task.

c. All crewmembers exit headfirst on hands and knees, proceeding carefully to ensure that no one is hurt.

d. The CDR is in command until the orbiter crew reaches the bunker, where the NTD is in command.

The table below details the Mode I sequence of action.

Table 5-2. Mode I Crew Egress

Step	Crewmember	Action
1		**Prep for egress**
	All	Upon call by CDR or NTD for emergency egress, immediately prepare for egress as follows (refer to emergency egress cue card): a. Check that neck dam tabs are released. b. Close and lock helmet visor. c. Turn on orbiter oxygen (to conserve EOS oxygen until quick disconnect from orbiter oxygen in substep j). d. Activate EOS by pulling green apple. e. Remove kneeboards and discard. f. Release quick disconnect for cooling. g. Release seat restraints by turning rotary buckle. h. Release parachute Frost fittings and ejector snaps.

Table 5-2. Mode I Crew Egress (continued)

Step	Crewmember	Action
		i. Release quick disconnects for comm.
		j. Release quick disconnects for Orbiter oxygen.
		k. Egress seat.
2		**Side hatch opening**
	MS3	**WARNING** Under no circumstances should the side hatch be jettisoned. If jettisoned, the side hatch would contact the launch tower and possibly destroy the OAA. a. Egress seat 5 carefully, using handhold strap on forward locker. b. Climb "down" to egress platform; proceed to side hatch. **CAUTION** Exercise care in pushing the hatch open. There is a 1-foot drop to the floor of the white room and a 6 in. to 18 in. gap between the orbiter and the edge of the white room. c. Rotate hatch handle to the VENT position, then rotate to the OPEN position (refer to decal). d. Open side hatch.
3		**Evacuate orbiter**
	All	**CAUTION** A portion of the right side of the white room folds back so that the OAA can move into position.

Table 5-2. Mode I Crew Egress (continued)

Step	Crewmember	Action
		This creates an unprotected area on that side: 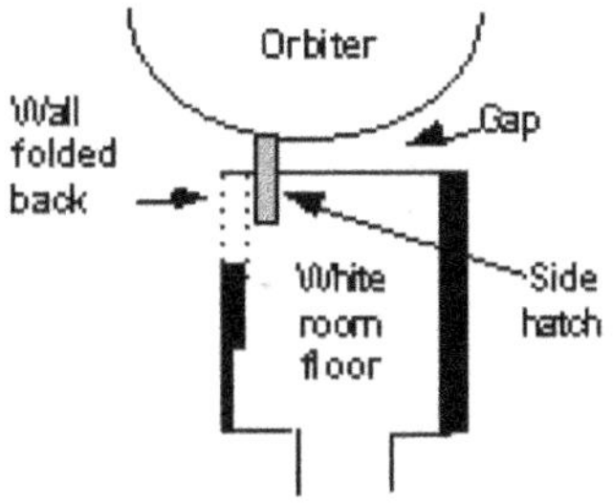 You can fall from this unprotected area if you veer too far right.
	All	Crawl out side hatch headfirst.
4		**Proceed to baskets**
	All	Follow marked route (shown below) through the white room and across the OAA and the FSS to the baskets (which are located in recesses on the west side of the FSS). The primary and alternate routes are as shown below: 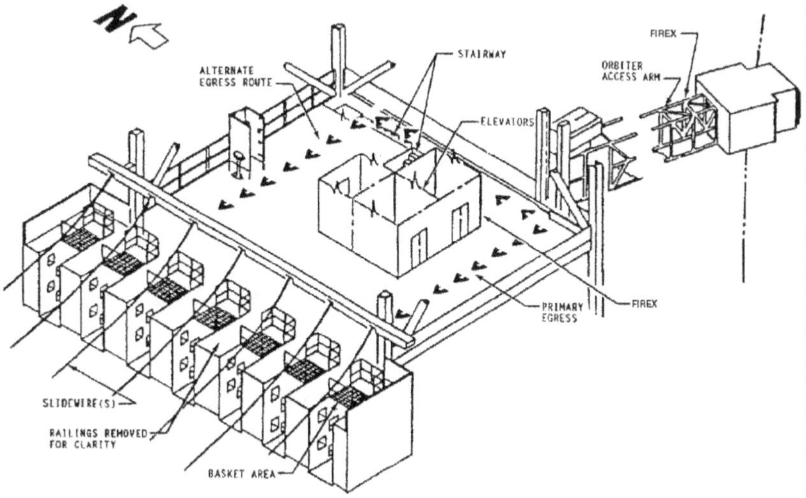 Figure 5-3. Primary and alternate routes to baskets

Table 5-2. Mode I Crew Egress (continued)

Step	Crewmember	Action
	All	a. <u>Primary route</u> is painted yellow with black chevrons pointing the direction. The primary route runs along the OAA, straight past the elevator doors, and on to the baskets. b. <u>Alternate route</u> is used if primary route is blocked. The secondary route also is painted yellow with black chevrons. It turns right, and then proceeds around the back of the elevators and on to the baskets. Note: Visibility enroute will be very poor because of water spray from the Firex fire extinguishing system and possibly because of smoke or vapor.
		Firex system activation
		Upon leaving the white room, crewmembers can activate the Firex fire extinguishing system if the system has not been activated by the Launch Control Center (LCC). The system sprays large amounts of water along the OAA and at the 195-foot level of the launch platform. Crewmembers activate the system by hitting two paddles just outside the white room on the right side of the OAA. The paddles, labeled "ARM" and "FLOW" (or "ACTUATE") can be pushed in any sequence. Crewmembers also can activate the Firex system immediately upon leaving the OAA. Additional activation paddles are located on the elevator shaft in front of the access arm.
5		**Descend in slidewire baskets**
	All	Descend as follows: a. Two to three persons enter each basket. b. Using the manually operated guillotine-type device (attached to the platform handrail), the crewmember nearest the guillotine in each group severs the rope securing the basket (a backup knife is stowed in the basket). c. All ride baskets to landing zone (takes 22 sec). d. After basket stops in landing zone, a crewmember in the basket pulls a D-ring to release the south side of the basket for easier egress.

Table 5-2. Mode I Crew Egress (concluded)

Step	Crewmember	Action
	All	**CAUTION** The baskets tend to rise as each crewmember exits. Exercise care in jumping out. **CAUTION** To avoid being hit by descending baskets, stay on the cement walkways enroute to the bunker. e. Exit basket and go directly to the bunker being sure to stay on cement walkways.
6		**In bunker**
	All	Follow NTD's directions whether to: a. Stay in bunker until rescuers arrive. The bunker has breathing air, water, phone line communications, and medical supplies. b. Evacuate in M-113 armored personnel carrier (Figure 5.4) to a designated helicopter landing site.

5.3.4 Armored Personnel Carrier

The M-113 armored personnel carrier used to evacuate the orbiter crew and fire rescue personnel from the pad area (Figure 5-4).

The M-113 is located at the bunker and driven to helipad sites considered safe for helicopters to land.

Figure 5-4. M-113 armored personnel carrier

5.3.5 Mode I Egress Cue Card Procedures

The emergency egress cue card is shown below with the Mode I pad egress procedures highlighted (Figure 5-5).

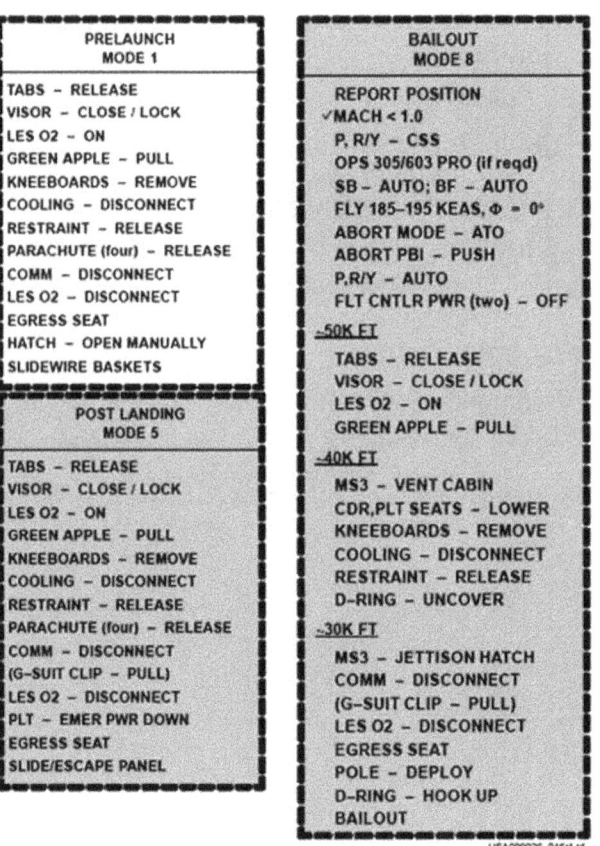

Figure 5-5. Mode I pad egress cue card procedures

The table below iterates the Mode I pad egress procedures in the left column as they appear on the emergency egress cue card (previous page) and gives an expanded version in the right column.

Table 5-3. Prelaunch

MODE I PAD EGRESS	
Cue Card Version	**Expanded Version**
TABS – RELEASE	Release neck dam tabs.
VISOR – CLOSE/LOCK	Close and lock helmet visor.
LES O2 – ON	Turn orbiter oxygen on.
GREEN APPLE – PULL	Activate emergency oxygen by pulling "green apple."
	Ensure green apple pulled full travel.
KNEEBOARDS – REMOVE	Remove kneeboards if worn.
COOLING – DISCONNECT	Release suit cooling quick disconnect.
RESTRAINT – RELEASE	Release seat restraints by turning rotary buckle.
PARACHUTE (four) – RELEASE	Release parachute from harness at shoulders and hips (four attachment points).
COMM – DISCONNECT	Release communications quick disconnect.
LES O2 – DISCONNECT	Release orbiter oxygen quick disconnect.
EGRESS SEAT	All crewmembers egress seats and move to side hatch.

Table 5-3. Prelaunch (concluded)

Mode I Pad Egress	
Cue Card Version	**Expanded Version**
HATCH – OPEN MANUALLY	a. All crewmembers evacuate orbiter. **CAUTION** Watch for gap between orbiter and white room and open wall. b. Go through white room and across OAA. **CAUTION** Water spray will reduce visibility and make footing slippery. Use buddy system.
SLIDEWIRE BASKETS	a. Get into slidewire baskets (three persons max. per basket, total seven baskets). b. Activate release mechanism. c. Ride basket down to landing area. **CAUTION** Stay on walkways after exiting baskets. d. Enter bunker and wait for instructions (may be instructed to leave area in M-113 armored personnel vehicle).

5.4 MODES II, III, IV – ASSISTED PAD EGRESS/ESCAPE

5.4.1 Introduction

In Modes II, III, and IV, the three <u>assisted</u> pad egress and escape modes, support personnel assist the flight crew in egressing from the orbiter and escaping to a safe place. Differences among these modes are discussed below. Egress following shutdown of the Space Shuttle Main Engine (SSME) also is discussed.

5.4.2 Egress Conditions

The conditions attending each of the assisted pad egress modes are as follows:

Mode II	Mode III	Mode IV
Seven-person closeout crew is still assisting flight crew with ingress (time from when flight crew enters the 195-foot level until closeout crew leaves).	Seven-person closeout crew has left the launch pad.	Seven-person closeout crew is still assisting flight crew with ingress.
Side hatch may be open or closed.	Side hatch is already closed.	Side hatch may be open or closed.
The emergency is such that members of the seven-person fire/rescue crew cannot be sent to the pad.	The emergency is such that members of the seven-person fire/rescue crew can be sent to the 195-foot level to rescue the flight crew.	The emergency is such that members of the seven-person fire/rescue crew can be sent to the 195-foot level to rescue the flight crew and the closeout crew.
Person in charge is the closeout leader until reaching the bunker, where the NTD is in charge.	Person in charge is the rescue team leader until reaching the bunker, where the NTD is in charge.	Person in charge is the rescue team leader until reaching the bunker, where the NTD is in charge.

5.4.3 Sequence of Action

The action taken by the flight crew and support personnel in the assisted pad egress modes is as follows:

Modes II, III, and IV

At call for mode egress, crewmembers immediately perform the following steps:

Step	Procedure
1	Close and lock visor.
2	Turn on orbiter O_2.
3	Activate EOS.
4	Pull quick disconnect for cooling.
5	Pull quick disconnect for comm.
6	Pull quick disconnect for orbiter O_2.

The action taken by the orbiter crew and support personnel in each of the assisted pad egress modes continues below:

Mode II	Mode III	Mode IV

Crewmembers egress assisted by the closeout crew.
 	
The closeout crew assists the flight crew as follows:
a. Assists crewmembers out of seats if necessary.
b. Ensures that crewmembers without helmets have oxygen.
c. Evacuates everyone to the slidewire baskets.
	If some crewmembers can egress on own, they:
	a. Open the side hatch.
	b. Assist others to the slidewire baskets.
	
	If no crewmember can egress on own, the fire/rescue crew:
	a. Opens the side hatch.
	b. Enters the orbiter.
	c. First checks:
	  1. Each crewmember's helmet closed.
	  2. EOS activated.
	d. Assists flight crew in escaping as follows:
	  1. If safe enough, brings crew down elevator.
	  2. If not safe enough, takes crew to slidewire baskets.
		The actions in Mode IV are the same as those in Mode III, except:
		a. The side hatch may still be open.
		b. Members of closeout crew can also assist others to slidewire baskets.
		c. There will be more persons needing assistance (flight crew and closeout crew).

The rest of the action taken is the same as that for Mode I:

a. Descend in baskets.

b. Go to bunker.

c. Follow NTD's direction whether to stay or evacuate in M-113.

5.4.4 SSME Shutdown

Following a shutdown of the SSME, the NTD receives information from LCC console operators about the status of the system and whether the system has shut down safely.

Once it is determined that the system has shut down safely, the orbiter crew is told that there is a launch scrub and to begin safing the orbiter systems from the flight deck.

Following SSME shutdown, the flight crew situation is as follows:

a. The flight crew does not know immediately whether there is an emergency, and the possibility exists that an emergency with subsequent mode egress could be declared at any time.

b. The flight crew is aware that SSME shutdown has occurred and that the crew will egress.

Crewmembers should immediately prepare to egress the orbiter as follows:

Note: Use the egress cue card as a guideline.

a. Keep visor down and locked with Orbiter O_2 – ON.

b. Remove all kneeboards.

c. Disconnect cooling.

d. Release seat restraint.

e. Release all four parachute attachments.

The flight crew will be then ready in either of the two following cases:

1. <u>If an emergency is declared and a mode egress called</u>, the flight crew will be ready to perform the following steps immediately:

Step	Procedure
1	Activate EOS by pulling "green apple." Note: Visor should already be closed and locked with Orbiter O_2 – ON.
2	Pull the quick disconnect for comm.
3	Pull the quick disconnect for orbiter O_2.
4	Egress orbiter.

2. <u>Even if no emergency is declared and the launch is scrubbed</u>, the flight crew will be that much closer to egressing once the closeout crew arrives. The flight crew should wait for clearance from NTD to turn off O_2 and raise visors.

5.5   MODE V – UNASSISTED POSTLANDING EGRESS/ESCAPE

5.5.1   Introduction

Orbiter crewmembers perform Mode V egress any time they have to leave the orbiter postlanding without assistance from convoy crew personnel (who assist the orbiter crew in escaping to a safe area from a fallback position). Refer to Figure 5-6 for Mode V postlanding egress cue card procedures and Figure 5-7 for convoy positioning.

There are three types of Mode V egress:

a. Hatch jettison Mode V egress, made when there is imminent danger to the orbiter or the crew; e.g., fire in the crew module.

b. Hatch on Mode V egress, made when egress has to be expedited because of a hazardous condition or when the landing occurs at a landing site that has no convoy crew support.

WARNING

If the hatch cannot be opened normally e.g., hatch is jammed, do not use side hatch jettison pyros. Rather then pushing the hatch away from the vehicle, the potential exists for the side hatch thrusters to become projectiles inside the crew module.

c. Window 8 Mode V egress, made when side hatch egress path cannot be used.

5.5.2 Egress Routes, Means

The egress routes and means for Mode V are as follows:

Primary Route – Side Hatch

The primary egress route is through the side hatch. Orbiter crewmembers use the escape slide to descend or use descent control devices ("sky genies") to lower themselves to the ground in case of slide failure.

Egress through the side hatch is faster, easier, and less hazardous. It is the egress route used unless:

a. The side hatch cannot be opened or jettisoned.

b. Hazardous conditions exist outside the side hatch.

Secondary Route – Escape Panel

The secondary egress route is through the Window 8 escape panel with the orbiter crew using sky genies to descend to the ground.

Since the side hatch can be jettisoned, there are only a few scenarios that would require using the Window 8 escape panel.

The table below summarizes the egress routes and the means used.

Table 5-4. Egress Route and Means Used

When egress is through...	Use...	
	As primary means:	As secondary means:
Side hatch (primary route) • Opened <u>normally</u> • Opened by jettisoning	**Slide**	**Sky genie** (If slide fails)
Window 8 escape panel (secondary route) • Opened <u>by jettisoning</u>	**Sky genie**	

5.5.3 Hatch-on Mode V Slide

Hatch-on Mode V slide egress procedures are as follows:

Step	Crewmember	Procedure
1		**Call for Mode V egress**
	CDR	Call for Mode V egress.
2		**Seat egress**
	All	At CDR call for Mode V egress, egress seat per cue card (for cue card, see "Mode V cue card procedures" in this section).
3		**Slide deploy**
	MS3	a. Unlock hatch handle and rotate handle counterclockwise to VENT position per hatch decal. b. Pull pin to release slide cover handle. c. Lift handle and remove slide cover. d. Lift girt bracket with the slide pack; lock bracket in place and rotate slide assembly ~270° into side hatch tunnel. e. Remove aft, then forward hinge pip pins. f. Lift and rotate entire slide pack back 90° and insert girt bracket into slide attach points (decals) on the hatch. g. Make sure slide is fully locked in place. h. Rotate hatch handle to the UNLATCHED position and push hatch open. **CAUTION** Side hatch is approximately 10 ft above the ground. i. Crawl out onto side hatch and push the slide off the outboard edge of the hatch. j. Pull the inflation lanyard all the way out **without releasing grip**.

Step	Crewmember	Procedure
4		**Hatch egress, slide descent**

	All	All crewmembers egress through the side hatch as follows:
		
		 Note: There is a two-person load limit on the side hatch (including person on the slide).
		
		 a. Crawl onto the side hatch and descend slide feet first, making sure to:
		   1. Keep legs and feet slightly apart and toes pointed up.
		   2. Keep hands at side of body to aid balance.
		   3. Lean back with weight off heels.
		   4. Be prepared to stand and move away on reaching bottom of slide.
		 b. Upon reaching bottom of slide:
		   1. Use forward momentum to assist in standing.
		   2. Move to one side.
		   3. Assist next person.

5		**Escape to safe area**
	All	Move crosswind to edge of runway, and then move upwind along edge of pavement. Note: Convoy crew personnel positioned in the fallback area (Figure 5-7) assist crewmembers as needed in escaping to a safe area.

5.5.4 Hatch-Jettison Mode V Slide Egress

The hatch-jettisoned Mode V slide egress procedures are as follows:

Step	Crewmember	Procedure
1		**Call for Mode V egress**
	CDR	Call for Mode V egress and advise ground personnel of intent to jettison side hatch.
2		**Seat egress**
	All	At CDR call for Mode V egress, egress seat per cue card (for cue card, see "Mode V cue card procedures" in this section).
3		**Hatch jettison**
	MS3	**WARNING** The cabin vent pyros are an ignition source in the payload bay and should never be activated postlanding. Note: MS3 removes pyro box safing pin before launch and entry. a. Open pyro box by squeezing latch knobs together and pulling cover down. b. Remove hatch jettison T-handle pip pin and white cloth cover. c. Squeeze and pull ribbed hatch jettison T-handle. Handle will come free from base. Note: To remove hatch jettison T-handle requires 20-50 lbs. of force.
4		**Slide deploy**
	MS3	a. Pull pin to release slide cover handle. b. Lift handle and remove slide cover. c. Rotate entire slide ~270° into hatch tunnel. d. Flip slide pack overboard. e. Pull inflation lanyard all the way out **without releasing grip**.

Step	Crewmember	Procedure
5		**Hatch egress, slide descent**

	All	a. Grab handhold above hatch tunnel; swing legs out and descend slide feet first, making sure to:
		    1. Keep legs and feet slightly apart and toes pointed up.
		    2. Keep hands at side of body to aid balance.
		    3. Lean back with weight off heels.
		    4. Be prepared to stand and move away on reaching bottom of slide.
		 b. Upon reaching bottom of slide:
		    1. Use forward momentum to assist in standing.
		    2. Move to one side.
		    3. Assist next person.

6		**Escape to safe area**
	All	Move crosswind to edge of runway, then move upwind along edge of pavement. Note: Convoy crew personnel positioned in the fallback area (Figure 5-7) assist crewmembers as needed in escaping to a safe area.

5.5.4.1 Side Hatch Egress with Sky Genie

Since the side hatch is the primary egress route, it is preferable in the event of egress slide failure to use the sky genies through the side hatch rather than through the Window 8 escape panel.

The procedures for sky genie use through the side hatch (hatch on or hatch off) are as follows:

Step	Crewmember	Procedure
1		**Sky genie deploy**
	Flight deck crewmember All	a. Remove sky genie and rope bag from stowage bag; free cable from cable track. b. Pass sky genie rope bag down to middeck. c. Manually pull down sky genie from flight deck for just enough slack to sit on hatch or in the side hatch tunnel if hatch jettisoned. d. Throw bag containing rope out hatch opening aft of hatch. e. Uncover and open carabiner and attach sky genie snap shackle to carabiner before exiting.

Step	Crewmember	Procedure
2		**Hatch egress, sky genie descent**
	All	Note: Distance to ground may be 10 ft or more. a. Crawl out on hatch or sit in side hatch tunnel. b. While sitting, make sure rope and cable are not tangled or caught on any part of suit or around arm or leg. c. Place feet on wing root, which is aft of side hatch. d. Lower down side of orbiter to ground; control descent rate by applying tension on rope. Note: Rescue team personnel also can control the descent rate by applying tension on the rope from the ground. **CAUTION** The orbiter fuselage curves under. Take care not to get too far out from the fuselage at the curvature. You could swing in and be injured. e. Upon reaching ground, disconnect snap shackle from carabiner (can open carabiner if snap shackle fails to release). f. Assist next person descending.
3		**Escape to safe area**
	All	Move crosswind to edge of runway, and then move upwind along edge of pavement. Note: Convoy crew personnel positioned in the fallback area (Figure 5-7) assist crewmembers as needed in escaping to a safe area.

USA009026
Basic

5.5.5 Window 8 Escape Panel Egress with Sky Genie

The procedures for egressing through the Window 8 escape panel using sky genies are as follows:

Step	Crewmember	Procedure
1		**Seat egress**
	All	Egress seat per cue card:
		┌─────────────────────────────────────┐ │ **ESCAPE PANEL EGRESS** │ │ CLEAR WINDOW │ │ ▩ TABS / VISOR / GREEN APPLE │ │ PULL PANEL JETT T-HANDLE │ │ PRY BAR / VENT IF REQD │ │ PLT EMERGENCY POWERDOWN │ │ DEPLOY & ATTACH SKY GENIE │ │ EGRESS CUE-1a/A, E/E │ └─────────────────────────────────────┘
2		**Panel jettison**
	CDR	Advise ground personnel of intent to jettison Window 8. Note: Fire personnel may direct a water fog across the top of the orbiter before panel jettison to dissipate any explosive/flammable atmosphere in the panel area.
	CDR/PLT	a. Remove Window 8 jettison handle pip pin and cover. b. Pull jettison handle. **WARNING** Do not jettison window until all flight deck crewmembers are breathing emergency O_2. Inner window severs and rotates inboard to aft console. Ensure that all crewmembers are clear of opening path.
	MS2	**WARNING** Window glass will shatter. Beware of falling glass around entire window area and down to middeck. If Window 8 does not open, use prybar (stowed in locker A17).

Step	Crewmember	Procedure
3		**Sky genie deploy**
	All	a. Remove sky genie and rope bag from stowage bag; free cable from cable track. b. Throw sky genie rope bag out Window 8 over starboard side of the orbiter. c. Uncover and open harness carabiner and attach sky genie snap shackle to carabiner before exiting.
4		**Escape panel egress, sky genie descent**
	All	**CAUTION** Take care not to drop the rope in the cabin. Dropping it could put knots in the rope or cause shattered glass and debris to stick to it, making descent difficult or impossible. Note: You do not need to hold on to sky genie when climbing up seat and out of opening. This allows use of both hands. a. Climb onto MS2's seat, egress step, and headrest, and then through Window 8. **WARNING** Crewmembers should stay clear of previous crewmember's descent cable/rope until rope is slack, signaling that the previous crewmember is on the ground. **WARNING** Following a nominal entry, the thermal protection system around the overhead window area can be 120° F 4 min after touchdown, and the overhead glass can be 190° F.

Step	Crewmember	Procedure
		b. While sitting on windowsill, make sure rope and cable are not tangled or caught on any part of suit or around arm or leg.
		c. Hold onto sky genie hand loop with left hand and onto loose end of rope with right hand.
		d. On your left side, slide down starboard side of orbiter until sky genie cable and rope are taut.
		e. Proceed to lower down starboard side of orbiter, controlling descent rate by applying tension on the rope.
		Note: Rescue team personnel also can control the descent rate by applying tension on the rope from the ground.
		f. Upon reaching ground, disconnect snap shackle from carabiner (can open carabiner if snap shackle fails to release).
		g. Assist next person descending.
5		**Escape to safe area**
	All	Move crosswind to edge of runway, and then move upwind along edge of pavement.
		Note: Convoy crew personnel positioned in the fallback area (Figure 5-7) assist crewmembers as needed in escaping to a safe area.

5.5.6 Mode V Cue Card Procedures

The emergency egress cue card is shown below with the Mode V postlanding egress procedures highlighted (Figure 5-5).

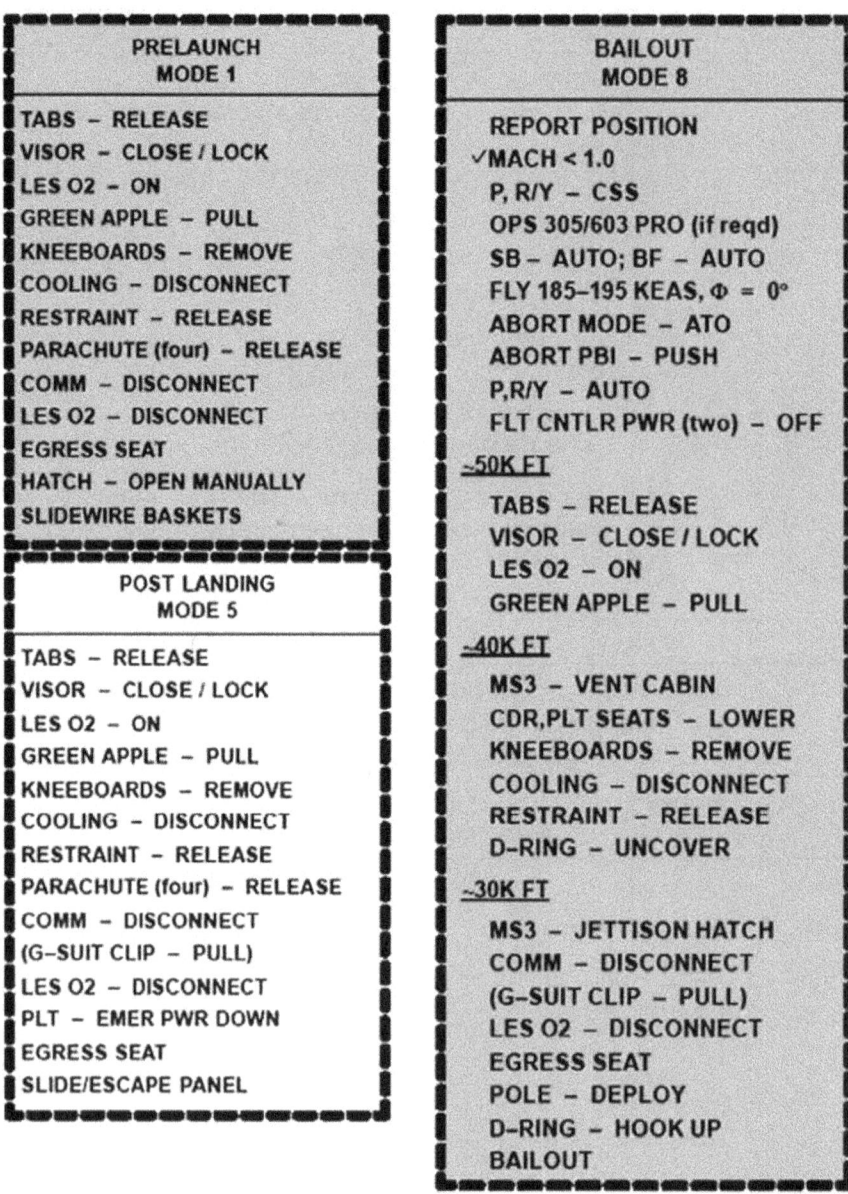

Figure 5-6. Mode V postlanding egress cue card procedures

The table below iterates the Mode V postlanding egress procedures in the left column as they appear on the emergency egress cue card and gives an expanded version of them in the right column.

Table 5-5. Postlanding

MODE V POSTLANDING EGRESS	
Cue Card Version	**Expanded Version**
TABS – RELEASE	Release neck dam tabs.
VISOR – CLOSE/LOCK	Close and lock helmet visor.
LES O2 – ON	Turn orbiter oxygen on.
GREEN APPLE – PULL	Activate emergency oxygen by pulling "green apple." Note: Ensure green apple pulled full travel.
KNEEBOARDS – REMOVE	Remove kneeboards if worn.
COOLING – DISCONNECT	Release suit cooling quick-disconnect.
RESTRAINT – RELEASE	Release seat restraints by turning rotary buckle.
PARACHUTE (four) – RELEASE	Release parachute from harness at shoulders and hips (four attachment points).
COMM – DISCONNECT	Release communications quick disconnect.
(G-SUIT CLIP – PULL)	Pull *g*-suit clip (only if returning from orbit after 2 or more days).
LES O2 – DISCONNECT	Release orbiter oxygen quick disconnect.
PLT – PWR DOWN	Pilot powers down per Emergency Pwrdn cue card.
EGRESS SEAT	All crewmembers egress seats and move to side hatch (unless Window 8 is used).
SLIDE/ESCAPE PANEL	Utilize slide/sky genie as described above.

5.5.7 Mode V Convoy Crew Positioning

Figure 5-7 below shows convoy crew positioning for Mode V (and Mode VI) postlanding crew egress.

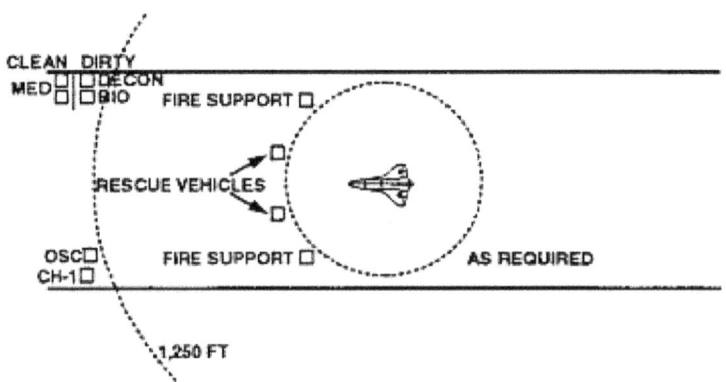

Figure 5-7. Convoy positioning, Mode V (and Mode VI)

5.5.8 Postlanding Loss of Orbiter Comm

In the event of orbiter comm loss postlanding, the crew can communicate with the convoy crew:

a. By using the PRC-112 survival radio.

b. If no radio contact, by using light signals.

The "Postlanding Procedures" section of the Entry Checklist contains information on each method. These two methods also are discussed below.

5.5.8.1 Using PRC-112 Survival Radio

If orbiter communication with the convoy crew and/or Mission Control Center (MCC) is lost, the CDR or PLT uses the PRC-112 survival radio (stowed in left suit leg pocket) as follows:

Step	Action
1	Select PRC-112 channel A (shuttle A/G channel programmed to 259.7 MHz) to establish voice comm with the convoy crew.
2	When comm has been established, the CDR informs the convoy crew of his/her intentions (Mode V, vehicle safing, powerdown, etc.).
3	If contact is not possible on channel A, select 243.0 MHz (UHF military emergency frequency).
4	Once contact is made, convoy commander changes to a nonemergency frequency.

5.5.8.2 Using Light Signals

If reestablishing comm with the PRC-112 is not successful, the flight crew can make its intentions known to the convoy crew by light signals made with a flashlight.

The flight crew continues signaling until the convoy crew acknowledges the signals by flashing the headlights on the convoy vehicle.

The table below defines the light signals used for specific actions.

Table 5-6. Light Signals

Signals from flight crew to convoy crew	
Action	**Signal**
Crew OK	Circular motion
OMS/RCS and side hatch safed	
Auxiliary Power Unit (APU) shutdown	
Crew needs assistance; will open/jettison hatch	Vertical motion
Crew needs assistance; will not open/jettison hatch	Horizontal motion
Signals from convoy crew to flight crew:	
The convoy crew acknowledges flight crew signals with a sequence of three flashes of the headlights on the convoy vehicle.	
Continuous flashes signal emergency powerdown and egress.	

5.6 MODES VI, VII – ASSISTED POSTLANDING EGRESS/ESCAPE

5.6.1 Introduction

In a Mode V postlanding situation, crewmembers egress from the orbiter and escape to a safe place on their own (convoy personnel assist from the convoy fallback position).

In a Mode VI or VII postlanding situation, support personnel assist the crew both in egressing from the orbiter and in escaping to a safe place.

These two modes are differentiated in Table 5-7.

Table 5-7. Description

Mode VI	Mode VII
The mishap occurs on or near a runway that is readily accessible to pre-positioned convoy ground personnel.	The mishap occurs off the runway within 25 n. mi. of the landing site.
Rescue personnel help the crew egress and escape.	Rescue and medical personnel are transported to the crash site by helicopter.

5.6.2 Mode VI Convoy Positioning

Mode VI convoy positioning is the same as that for Mode V (Figure 5-7).

5.7 MODE VIII – BAILOUT

5.7.1 Introduction

Mode VIII emergency egress makes it possible for the orbiter crew to bailout safely from the orbiter during controlled gliding flight at an altitude of 30,000 ft or less.

Bailout could occur during:

a. Any launch abort (Return-to-Launch-Site (RTLS), Transoceanic Abort Landing (TAL), Abort Once Around (AOA)).

b. Emergency deorbit that must be made regardless of landing site.

c. Any failure precluding a normal landing on a runway.

5.7.2 Bailout Sequence of Action

The table below details the bailout sequence:

Table 5-8. Mode VIII Crew Egress

Alt. (ft)	Step	Crewmember	Action
50,000	**1**		**Call for bailout**
		CDR	If range to go exceeds ~ 55 n. mi. at altitude of 50,000 ft: a. Call for bailout. b. Place orbiter in minimum sink rate attitude. c. Engage autopilot.
		All	At CDR call for bailout: a. Lower and lock helmet visor. b. Suit O_2 – ON (to breathe orbiter O_2 until quick disconnect, when emergency O_2 starts flowing). c. Pull green apple.
40,000	**2**		**Cabin venting**
		CDR	Request MS3 to vent cabin.

		MS3	At CDR request to vent cabin:
			a. Vent cabin as follows:
			  1. Open pyro box.
			  2. Pull safing pin on cabin vent T-handle (aft handle).
			  3. Squeeze and pull smooth pyro vent T-handle (forward handle).
			b. During cabin vent, monitor altimeter on middeck locker and communicate reading to CDR.

		CDR	During cabin vent, monitor altimeter reading and communicate reading to MS3.

Table 5-8. Mode VIII Crew Egress (continued)

Alt. (ft)	Step	Crewmember	Action
~31,000			**Cabin pressure equalization**
			Note: Altimeter reading rises as cabin pressure decreases. In ~75 sec (~31,000 ft) altimeter peaks.
	3		**Seat egress prep**
		All	**WARNING** Take care not to disconnect parachute inadvertently during egress prep.
			During cabin vent: a. Remove kneeboards. b. Break cooling and any biomed connections. c. Release seat restraints by turning rotary buckle. d. Remove cover to parachute D-ring on right riser and extend D-ring.
~30,000	4		**Side hatch jettison**
		CDR	When pressure equalized, request MS3 to jettison side hatch.
		MS3	At CDR request to jettison side hatch: a. Pull safing pin on hatch jettison T-handle. b. Squeeze and pull ribbed hatch jettison T-handle (forward handle).
		All	c. If bailout after 2 or more days in space, pull *g*-suit clip. d. Disconnect comm and orbiter O_2.

Table 5-8. Mode VIII Crew Egress (concluded)

Alt. (ft)	Step	Crewmember	Action
	5		**Escape pole deploy**

		MS3	a. Egress seat and fold down seat back.
			b. Follow decal to deploy escape pole.
			If pole does not deploy fully (green stripe not visible), deploy pole manually:
			  1. Pull manual backup pip pin.
			  2. Pump ratchet handle in direction of pole (power stroke is upward) until green stripe visible.

		All	Egress seat and move to side hatch.
~ 30,000	6		**Bailout**

		All	At side hatch:
			  1. Attach D-ring on parachute riser to snap hook on lanyard (before releasing snap hook from magazine).
			  2. Release lanyard from magazine.
			  3. Kneel into hatch.
			  4. Bailout. Remain tucked and forcibly roll out of hatch.

5.7.3 Mode VIII Bailout Cue Card Procedures

The emergency egress cue card is shown below with the Mode VIII Bailout egress procedures highlighted (Figure 5-5).

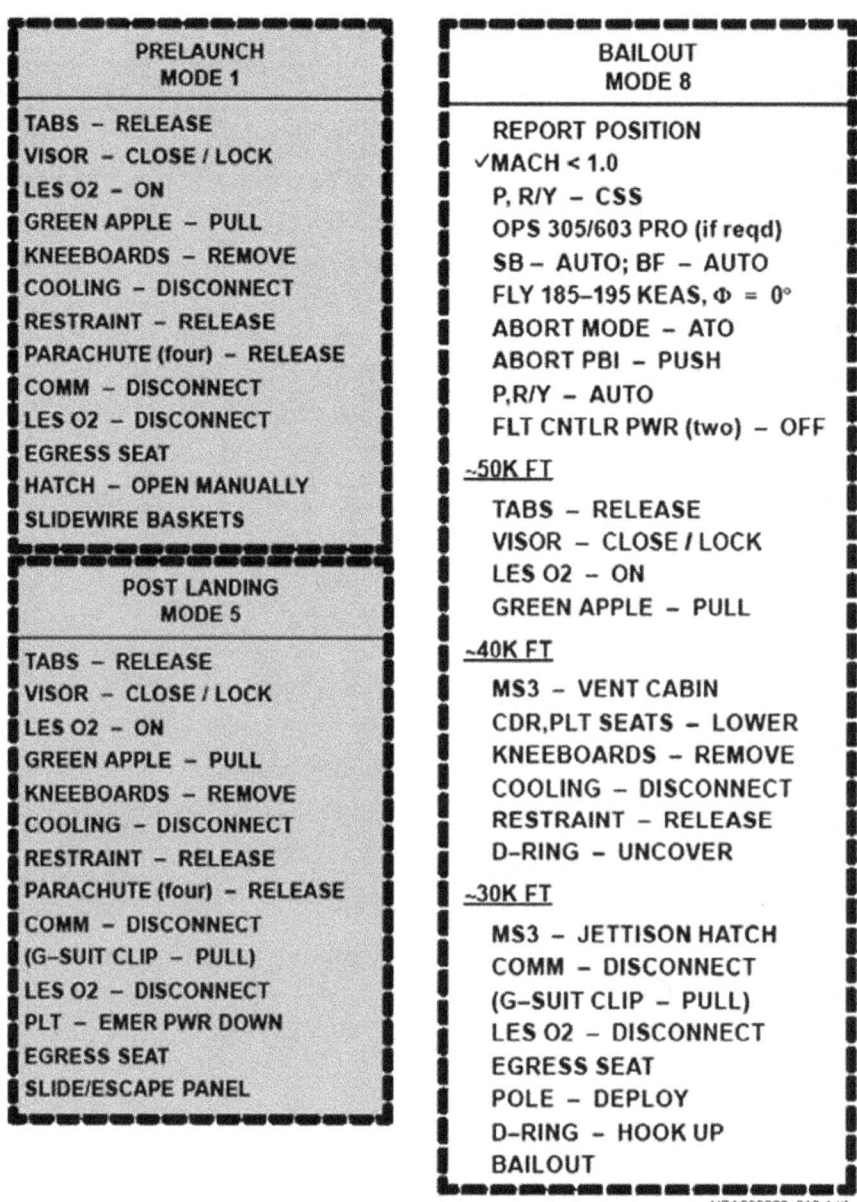

Figure 5-8. Mode VIII bailout cue card procedures

The table below iterates Mode VIII bailout procedures in the left column, as they appear on the emergency egress cue card, and gives an expanded version in the right column.

Table 5-9. Bailout

MODE VIII BAILOUT	
Cue Card Version	**Expanded Version**
REPORT POSITION	Contact Houston with TACAN radial and DME, altitude, heading, and indicated airspeed.
✓**MACH <1.0**	Verify Mach less than 1.0.
P,R/Y – CSS	Press CSS pushbutton indicators for both pitch and roll/yaw (enables rotational hand controllers for manual control).

OPS 305/603 PRO (if reqd)	Using PASS keyboard:
	    If end of mission or TAL or ELS:
	        Type "OPS 305 PRO"
	    If RTLS:
	        Type "OPS 603 PRO"

SB - AUTO; BF – AUTO	Press AUTO pushbutton indicators for both speed brake and body flap.
FLY 185-195 KEAS, Φ = 0°	Stabilize airspeed between 185 and 195 with bank angle of 0°.
ABORT MODE – ATO	Turn rotary ABORT MODE dial to ATO (abort to orbit).
ABORT PBI - PUSH	Press ABORT pushbutton indicator.
P, R/Y – AUTO	Press AUTO pushbutton indicators for both pitch and roll/yaw.
FLT CNTLR PWR (two) – OFF	Flip both FLT CONTLR PWR toggle switches to OFF (disables rotational hand controllers to prevent accidental inputs when egressing seat).
~50K FT	
TABS – RELEASE	Release neck dam tabs.
VISOR – CLOSE/LOCK	Close and lock helmet visor.
LES O2 – ON	Turn orbiter oxygen on.
GREEN APPLE – PULL	Activate emergency oxygen by pulling "green apple." Note: Ensure green apple pulled full travel.

Table 5-9. Bailout. (continued)

Cue Card Version	Expanded Version
~40K FT	
MS3 – VENT CABIN	CDR gives command to MS3 to vent cabin by pulling cabin vent T-handle.
CDR, PLT SEATS – LOWER	CDR and PLT lower seats.
KNEEBOARDS – REMOVE	Remove kneeboards if worn.
COOLING DISCONNECT	Release cooling quick disconnect.
RESTRAINT – RELEASE	Release seat restraints by turning rotary buckle.
D-RING – UNCOVER	Expose and extend D-ring and bridle for pole hookup on right parachute riser.
~30K FT	
MS3 – JETTISON HATCH	CDR gives command to MS3 to jettison side hatch by pulling hatch jettison T-handle.
COMM - DISCONNECT	Release communications quick disconnect.
(G-SUIT CLIP – PULL)	Pull *g*-suit clip (only if returning from orbit after 2 or more days).
LES O2 – DISCONNECT	Release orbiter oxygen quick disconnect.
EGRESS SEAT	All crewmembers leave seats and move to side hatch.
POLE – DEPLOY	MS3 deploys pole per decal.
D-RING – HOOK UP	Hook D-ring to snap hook on pole.
BAILOUT	At hatch: a. Pull snap hook from pole magazine. b. Kneel into hatch. c. Bailout. Remain tucked and forcibly roll out of hatch.

5.7.4 Parachute Deploy Sequence

The table below gives the time/altitude and action during the parachute deployment sequence.

Table 5-10. Deployment Sequence

Time/alt (ft)	Step	Crewmember	Action
	1		**Parachute deploy activation**
		All	Normally, the parachute deployment sequence is activated automatically, but should auto activation fail, activate sequence manually. a. <u>Auto activation</u>:
T = 0.0 sec			1. As each crewmember exits the orbiter, the windstream must exert a minimum of 150-lb force on the crewmember and consequently on the bridle.
T = 1.5 sec			2. A pyrotechnic cutter separates the bridle and the lanyard (by this time crewmember is off the escape pole and away from the orbiter).
T = 3.0 sec			3. The 18-in. pilot chute deploys. The pilot chute will deploy the 4.5-foot drogue chute. The drogue chute will: (a) Stabilize the crewmember down to 14,000 ft. (b) Arms the AAD, which allows the drogue chute to deploy the main canopy. b. <u>Manual activation</u> (if auto activation fails): 1. Pull the ripcord on the left parachute riser (to activate a single pyrotechnic cutter that deploys the pilot chute after a 2 sec delay). 2. The sequence of events is identical to the auto opening sequence.

Table 5-10. Deployment Sequence (concluded)

Time/alt (ft)	Step	Crewmember	Action
~14,000	2		**Main canopy deployment**

		All	Main canopy deployment is activated automatically, but should the AAD fail, activate main canopy deployment manually.
			a. Auto deploy:
			   1. When the AAD senses a pressure altitude of 14,000 ± 1,000 ft, a pyrotechnic charge fires, allowing the drogue chute to deploy the 26 foot main canopy. During main canopy deployment the:
			SARSAT is activated.
			Life raft compartment is loosened.
			   2. The main canopy is reefed for 2 sec slowing the descent from ~120 ft/sec (freefall) to 80 ft/sec.
			   3. After 2 sec reefing:
			      (a) Redundant cutters cut the reefing line, allowing the main canopy to open fully.
			b. Manual deployment (if AAD fails):
			   1. Pull the drogue chute release knob (red knob) on the left parachute riser (this allows the drogue chute to immediately deploy the main canopy).
			Note: The drogue chute release knob will function only if the drogue chute has been deployed.

	3		**LPU activation (if water landing)**
		All	Prior to a water landing: a. Pull the manual inflation toggles for the LPU (an additional safeguard, since the LPUs are designed to inflate automatically upon immersion in water).
	4		**Landing**
		All	See next section.

5.7.5 Sequence of Action When Landing in Water

The table below describes the sequence of action when landing in water.

Table 5-11. Water Landing

Step	Action
1	When immersed in salt water: a. SEAWARS automatically releases the crewmember from the parachute risers. b. Life raft automatically deploys and partially inflates.
2	Before ingressing raft: a. Locate sea anchor/wicking sea dye and throw out of raft (dye begins wicking green). b. Locate bailing cup/pump and nonwicking sea dye and throw out of raft.
3	Move to foot of raft and float survival water packet in front of self.
4	Release one parachute ejector snap and bring remainder of parachute pack to front of self. Note: Since the raft is tethered to the parachute pack, release only one ejector snap to ensure that you stay connected to the raft at all times.
5	Ingress raft as follows: a. Grasp and roll back spray shield. b. Sink foot of raft behind self. c. Flood raft underneath self while lying back. d. Pull raft under self. e. Kick feet slowly to remain upright in water; be careful not to get tangled in tethers.

Table 5-11. Water Landing (continued)

Step	Action

6	After entering raft, proceed as follows:
	a. Pull spray shield up and over shoulders.
	b. Inflate second volume of the raft by pulling the lanyard on the upper CO_2 cylinder on the right side of raft (inflating the second volume raises the raft higher in the water and makes it more stable).
	c. Bail water out of the raft.
	   1. If the sea is relatively calm, remove most of the water with the spray shield and the bailing cup. Seal the spray shield over body (take care to seal the Velcro in such a way that water will not leak in). Pump out the remaining water with the bailing pump.
	   2. If the sea is rough, seal the spray shield over body and pump out the water with the bailing pump.
	Note: Getting the raft dry and keeping it dry are important for survival in cold conditions.
	
	Do not remove the harness at any time. The raft, emergency water, and the LPU are connected to the harness. Should you fall out of the raft, you will not be separated from the raft, and the LPU will provide buoyancy.
	
	Keep the helmet on at all times. The helmet is not tethered and protects the head during rescue operations.

7	Unpack and configure SARSAT signaling equipment stowed in pocket on left spray shield (beacon should remain in pocket unless foul weather dictates that it be brought inside): a. Extend rigid antenna fully (otherwise, signal strength will be diminished). Note: If rigid antenna does not extend, leave flexible antenna in place. b. After extending rigid antenna, disconnect flexible antenna.
8	When an aircraft or ship is seen or heard, try to make contact via voice on the PRC-112 survival radio (stowed in suit right lower leg pocket). **CAUTION** Do not deplete smoke/flare if no visual or audio contact is made with SAR forces. Depending upon how far bailout was from land, it could take 8 hours or more before rescue forces arrive. The first rescue force to arrive will probably be personnel in a C-130 transport aircraft specially equipped for search and rescue.

5.7.6 Contingency Landing Site Resources

The table below describes the resources available at each contingency landing site.

Table 5-12. Contingency Landing Sites and Resources

Site	Resources
East coast abort landing sites	Designated east coast DOD bases dedicate a runway to serve as an abort landing site (as an alternative to bailout if two or three shuttle main engines fail during ascent prior to TAL abort capability).
TAL	The TAL primary site and secondary site both have a C-130 in place at L - 24 hr, each with three 3-member PJ teams and survival equipment for SAR/MEDEVAC support. A MEDEVAC C-130 is on alert at Moron AB, Spain. The C-130 crews are on cockpit alert from L - 30 min until the TAL opportunity passes.
AOA	Helicopter support forces are at the AOA runway at landing (Ld) – 30 min.
Early termination landing	Helicopter support forces are at the runway at Ld-30 min for: a. Scheduled daily Primary Landing Site (PLS) opportunities when landing is declared. b. Unscheduled PLS landing (notification time permitting). c. Emergency Landing Site (ELS) (notification time permitting).

5.7.7 Launch Contingency Bailout Area Rescue

Department of Defense (DOD) SAR forces are postured to respond in the event of a launch emergency. In the event of a bailout, the SAR response is as follows:

- If bailout is within 200 n. mi. of KSC, crewmembers will be located by personnel in a C-130 transport aircraft specially equipped for search and rescue and picked up by the rescue helicopters (Jolly's) and transported to area hospitals.

- If bailout is greater then 200 n. mi. from KSC, crewmembers will be located by the C-130 aircraft. Two teams, each consisting of three Pararescue Jumpers (PJ) with a motorized rescue boat, will drop from the C-130 and pick up individual crewmembers as they work toward the middle of the "survivor string." The C-130 will also drop two 20-person life rafts for the assembled crewmembers and rescue personnel to wait in until rescue helicopters arrive. The helicopters will pick up the waiting crewmembers and personnel and transport the crewmembers to area hospitals.

The tables below outline SAR response for the KSC area and ECAL/TAL.

Table 5-13. KSC Area SAR Response

Resources	Response Time
Launch and RTLS (KSC) • Helicopters (one for every two crewmembers) at KSC Shuttle Landing Facility (SLF) configured for SAR and Medical Evacuation (MEDEVAC). Launch and RTLS (downrange) • The helicopters above refueled by H/KC-130 tanker aircraft for downrange up to 200 n. mi. • One each HC-130 and KC-130 aircraft at Patrick AFB (one spare) with two 3-man teams of PJ's and survival equipment; three rescue packages on helicopters; one HC-130 aircraft 175 - 200 n. mi. downrange L - 10 min with PJ's and rescue package. • One E-2C aircraft on alert at Patrick AFB and one on alert as backup at Naval Air Station (NAS) Norfolk to assist in aircraft control and communications relay. • U.S. Navy ship with helo at 150 n. mi. • Coast Guard cutter with helo at 100 n. mi. • Assistance from Solid Rocket Booster (SRB) recovery ships.	SAR forces can: • Locate and assist orbiter crew within 3 hours of notification. • Recover orbiter crew within 6 hours.

Table 5-14. ECAL/TAL Area SAR Response

Resources	Response Time
• HC-130 aircraft, CONUS. • P-3C aircraft on alert, ready to launch within 30 min from selected locations, drop SAR kits, provide surveillance. • C-130 aircraft from KSC with two 3-man PJ teams or from primary and secondary TAL sites with three 3-man PJ teams (C-130 MEDEVAC aircraft on alert at Moron AB, Spain). • U.S. Navy vessels, USCG cutters and aircraft, ships of opportunity used for recovery.	• SAR forces can: • Locate and assist orbiter crew within 24 hours of notification. • Recover orbiter crew as soon as possible.

All night rescue operations will be "best effort" by SAR forces.

The recovery posture for the launch contingency bailout area is illustrated below (Figure 5-9, and Figure 5-10).

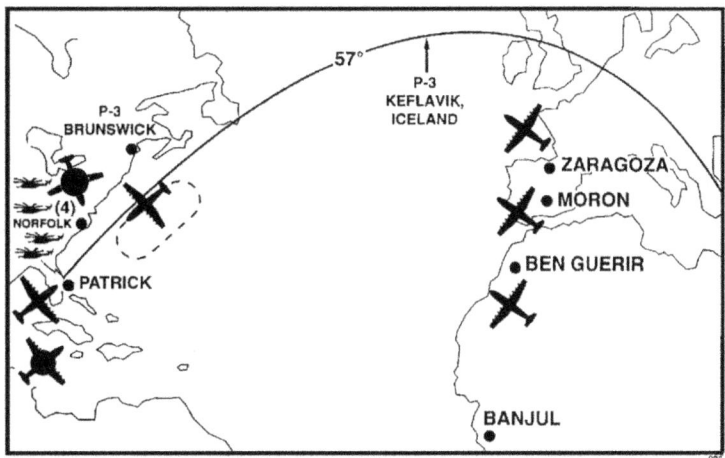

Figure 5-9. SAR Recovery Posture

The SAR recovery posture for the launch contingency bailout area is illustrated below (Figure 5-10).

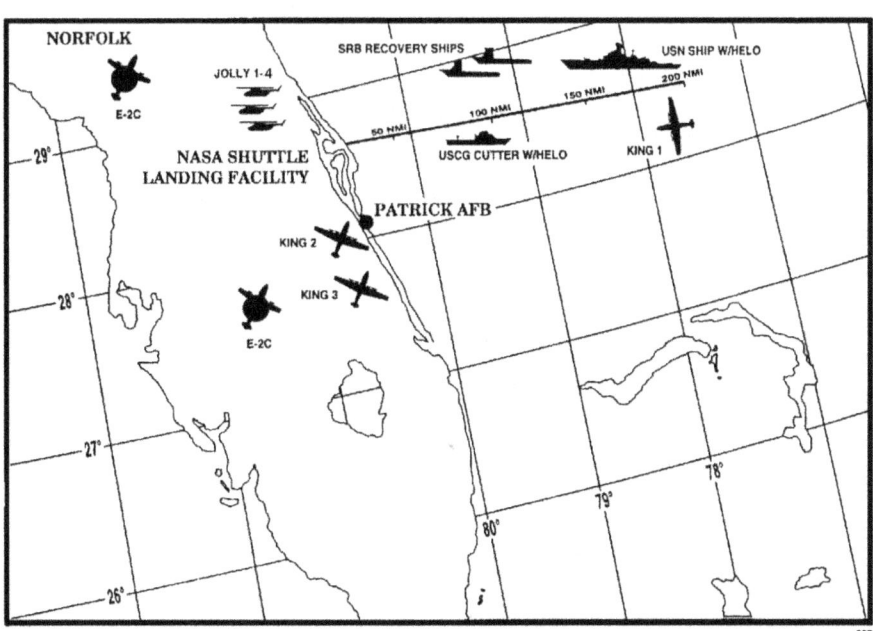

Figure 5-10. KSC area SAR recovery posture

APPENDIX A
ACRONYMS AND ABBREVIATIONS

A/A	Air to Air
A/G	Air to Ground
AAD	Automatic Actuation Device
ACES	Advanced Crew Escape Suit
AFB	Air Force Base
AOA	Abort Once Around
APU	Auxiliary Power Unit
ASP	Astronaut Support Personnel
ATU	Audio Terminal Unit
B/U	Backup
BIP	Bioinstrumentation Passthrough
CCA	Communications Carrier Assembly
CCCD	Crew Compartment Configuration Drawing
CCW	counterclockwise
CDR	Commander
CO_2	carbon dioxide
CONUS	Continental United States
CSS	Control Stick Steering
CW	clockwise
dc	direct current
DME	Distance Measuring Equipment
DOD	Department of Defense
ECAL	East Coast Abort Landing
EESS	Emergency Egress Slide System
ELS	Emergency Landing Site
EOM	End of Mission
EOS	Emergency Oxygen System
FD	Fight Director
FDF	Fight Data File
FSS	Fixed Service Structure
g	gravity
HIU	Headset Interface Unit
icom	intercom
ICU	Individual Cooling Unit
KSC	Kennedy Space Center

LCC	Launch Control Center
LCG	Liquid Cooling Garment
Ld	landing
LPU	Life Preserver Unit
LWS	Lightweight Seat
MAR	Middeck Accommodations Rack
MCC	Mission Control Center
MEDEVAC	Medical Evacuation
MS	Mission Specialist
n. mi.	nautical mile
NAS	Naval Air Station
NASA	National Aeronautics and Space Administration
NTD	NASA Test Director
O_2	oxygen
O&C	Operations and Checkout
OAA	Orbiter Access Arm
PBI	Push Button Indicator
PJ	Pararescue Jumper
PLS	Primary Landing Site
PLT	Pilot
PPA	Personal Parachute Assembly
PS	Payload Specialist
PSCS	Personal Suit Cooling System
psi	pounds per square inch
psia	pounds per square inch absolute
psig	pounds per square inch gauge
PTT	Push to Talk
QD	Quick Disconnect
RHC	Rotational Hand Controller
RSK	Recumbent Seat Kit
RTLS	Return to Launch Site
S	seat
SAR	Search and Rescue
SARSAT	Search and Rescue Satellite-Aided Tracking
SEAWARS	Seawater Activated Release System
SLF	Shuttle Landing Facility
SRB	Solid Rocket Booster
SSME	Space Shuttle Main Engine

T	time
TACAN	Tactical Air Navigation
TAL	Transatlantic Abort Landing
TELCU	Thermal Electric Liquid Cooling Unit
UHF	Ultrahigh Frequency
USCG	United States Coast Guard
USN	United States Navy
V dc	volts dc
VLHS	Very Lightweight Headset
w/helo	with helicopter
xmit	transmit

APPENDIX B
CREW ESCAPE LESSONS

1. <u>Escape Equipment and Systems 21001</u> - This lesson is an introduction to the various equipment and systems associated with crew escape from the orbiter.

   <u>Section 1</u> - Crew worn equipment including the ACES, harness, parachute and survival gear

   <u>Section 2</u> - Orbiter emergency egress equipment and systems including cabin vent, hatch jettison, Window 8 escape panel, escape pole, egress slide and sky genie.

   <u>Section 3</u> - Emergency egress equipment and systems located at the launch pad. This includes the white room, Orbiter access arm, slide wire baskets, bunker and M-113 personnel carrier.

   <u>Section 4</u> - Familiarization of the Orbiter's emergency egress equipment in the trainer. In addition, the students will be able to deploy the escape pole trainer.

2. <u>Escape Procedures 21001</u> - This classroom lesson introduces the student to the various procedures associated with emergency egress from the Orbiter. The eight emergency egress modes are described. Off-nominal conditions such as SSME shutdown, Launch Scrub, and Break up/Loss of control are also touched upon. After this, the procedures for the three unassisted modes are discussed. The Emergency Egress Cue Card is used to acquaint the student with the procedures. Following the discussion, the Escape Procedures video is shown.

3. <u>Escape Procedures 21020</u> – This lesson covers all procedures for a Mode V and Mode VIII escape. After donning the ACES, crewmembers will be strapped in and all aspects of a Mode VIII bailout will be practiced, including cabin vent (simulated), side hatch jettison (simulated), seat egress, pole deployment, pole hook-up and bailout (simulated). The student will then practice using the sky genie while suspended under the overhead crane. Then the student will be strapped into the FFT and the crew will do a Mode V egress out Window 8 using the sky genies. Once the entire crew is out and the FFT is ready, the crew will sit in their seats (no strap in) and perform a Mode V out the side hatch, again using the sky genies. Last, the students will reingress the FFT and will perform a Mode V out the side hatch using the slide.

4. <u>Reach and Vis 21020</u> - This lesson is designed to familiarize the student with Orbiter ingress and strap-in procedures. Once strapped in, the student is given the opportunity to explore their reach and visibility envelopes, first with an unpressurized suit, then again with a pressurized suit. The instructor will give suggestions on how to increase the envelope. The Suit Techs may provide different sized helmet pads and headrest pads to get each crewmember's preference. At the end of the session, a Mode I egress is performed.

5. <u>Water Survival 21001</u> - In this classroom session the helmet, ACES, parachute, harness, raft and survival equipment are discussed in detail as related to water survival and bailout. Inflation of the LPU and life raft are performed. A full complement of survival gear will be available for the students to see and handle. This class has a currency requirement of 2 years.

6. <u>Water Survival 21027</u> - This lesson enables the student to experience bailout, water entry, and water survival techniques. Each suited student will be suspended by parachute risers above the pool and they will perform post-parachute opening procedures, including inflating their life preserver unit (LPU). The student will be released from the hoist and they will enter the water. The risers will be manually released by the student. The student will extricate himself from underneath the parachute canopy. The student will perform a back drag and a front drag to practice proper body positioning and Frost fitting release for a parachute drag. The student will enter the life raft and bail water out of the raft, access all the survival gear and "use" it. The student will then attempt to re-ingress the fully inflated raft. Once complete with the raft exercises, the student will perform the procedures associated with the Forest Penetrator rescue seat. Once this is complete, the student will perform the side hatch bailout. This class has a currency requirement of 2 years.

7. <u>Postlanding Egress 91019</u> - After donning the ACES the student will practice using the sky genie while suspended under the overhead crane. Then the student will be strapped into his ascent seat in the FFT and the crew will do a Mode V egress out Window 8 using the sky genies. Once the entire crew is out and the FFT is ready, the crew will get into their entry seating positions and perform a Mode V out the side hatch, again using the sky genies. Students will perform a Mode V out the side hatch using the slide.

8. <u>Bailout 91020</u> - This lesson provides more practice on procedures used for in-flight bailout, emphasizing crew coordination. All aspects of the bailout will be practiced, including cabin vent (simulated), side hatch jettison (simulated), seat egress, pole deployment, pole hook-up and bailout (simulated). Two runs will be performed, one in ascent seating and the other in entry seating. During the second run, the g-suits will be inflated. Additionally, the prime and backup pole operators will perform pole deployment using the engineering egress pole trainer.

9. <u>Prelaunch Procedures 91020</u> - This session is a dry-run of the TCDT prelaunch activities. It includes the activities the ASP does prior to crew ingress (FDF set-up, cabin prep, etc).

10. <u>Prelaunch Ingress/Egress 91020</u> - This class is performed in conjunction with Prelaunch Procedures 91020 in which the ASP practices configuring the Orbiter for crew ingress. With the CCT/CCTII in the vertical position, students perform nominal ingress procedures assisted by the Astronaut Support Personnel (ASP) and suit technicians. Included in this lesson are the crew, NASA Test Director (NTD), and Orbiter Test Conductor (OTC), and MCC comm checks. A Mode I egress is performed allowing the crew to practice an unaided egress on the pad.

11. <u>Escape Refresher 41020</u> - This classroom lesson reviews the various procedures associated with emergency egress from the Orbiter. Off-nominal conditions such as SSME shutdown, Launch scrub, and Break up/Loss of Control are also touched on. The procedures for the three unassisted modes are discussed. The Emergency Egress Cue Card is used to review the procedures. The Escape Procedures video is shown. Prybar ops are reviewed in the CCT.

DOCUMENT NUMBER: USA009026
Basic

TITLE: Crew Escape Systems 21002

NASA-JSC

AC	Engle, J.
CA4	Hanley, R. (6)
CB	Rominger, K
	Ivins, M
DA8	DA8 Library
DF52	Jason, J.
	Shimp, J.
DX	Prince, A.
DX3	Crew Escape (25)
EC	Ellis, W.
EC5	Chhipwadia, K. (10)
EP5	Brown, C.
MV	Wood, D.
MV5	Sauser, B.
SD2	Billica, R (10)

NASA-KSC

USK-170	Arnold, P. (3)
USK-155	Parks, B.
USK-155	Clark, K

BOEING/DOWNEY

AD60	Shartz, I. (4)

UNITED SPACE ALLIANCE

H10H-115	HEI-SFOC
USH-121G	SFOC Technical Library (2)

DOD MANNED SPACE FLIGHT SUPPORT OFFICE

1201 Edward H. White II St. (10)
Patrick AFB, FL 32925-3239
(84)

National Aeronautics and Space Administration

Space Shuttle Transoceanic Abort Landing (TAL) Sites

NASAfacts

Space Shuttle Discovery blasts off from Launch Pad 39B at dawn March 8, 2001, on mission STS-102, the eighth flight to the International Space Station.

Planning for each space shuttle mission includes provisions for an unscheduled landing at contingency landing sites in the United States and overseas. Several unscheduled landing scenarios are possible, ranging from adverse weather conditions at the primary and secondary landing sites to mechanical problems during the ascent and mission phases that would require emergency return of the orbiter and its crew.

Types of Unscheduled Landings

The Transoceanic Abort Landing (TAL) is one mode of an unscheduled landing. The orbiter could have to make an unscheduled landing if one or more of its three main engines failed during ascent into orbit, or if a failure of a major orbiter system, such as the cooling or cabin pressurization systems, precluded satisfactory continuation of the mission.

Several unscheduled landing scenarios are possible, with available abort modes that include Return to Launch Site, Launch Abort Site landing, Transoceanic Abort Landing (TAL), Abort Once Around, and Abort to Orbit. The abort mode would depend on when in the ascent phase an abort became necessary.

The TAL abort mode was developed to improve the options available if failure occurred after the last opportunity for a safe Return To Launch Site or Launch Abort Site landing, but before the Abort Once Around option became available. A TAL would be declared between roughly T+ 2:30 minutes (liftoff plus 2 minutes, 30 seconds) and T+ 7:30 minutes. Main engine cutoff occurs about T+ 8:30 minutes into flight, with the exact time depending on the payload and mission profile.

A TAL would be made at one of three designated sites: Istres Air Base in France, Zaragoza Air Base in Spain and Moron Air Base in Spain.

Each TAL site is covered by a separate international agreement. The TAL sites are referred to as augmented sites because they are equipped with space shuttle-unique landing aids and are staffed with NASA, contractor and Department of Defense personnel during a launch and contingency landing.

Space shuttles are launched eastward over the Atlantic Ocean from Kennedy Space Center in Florida for insertion into low- to high-inclination orbits. Depending on mission requirements, an orbiter follows an orbital insertion inclination between 28.5 degrees (low) and 57.0 degrees (high) to the equator. All space shuttle launches to the International Space Station use an inclination of 51.6 degrees. The lower inclination launches allowed for a higher maximum payload weight but are no longer used.

High- or low-inclination launches require different contingency landing sites, with two or three of the landing sites staffed to ensure there is acceptable weather for a safe landing at a TAL site.

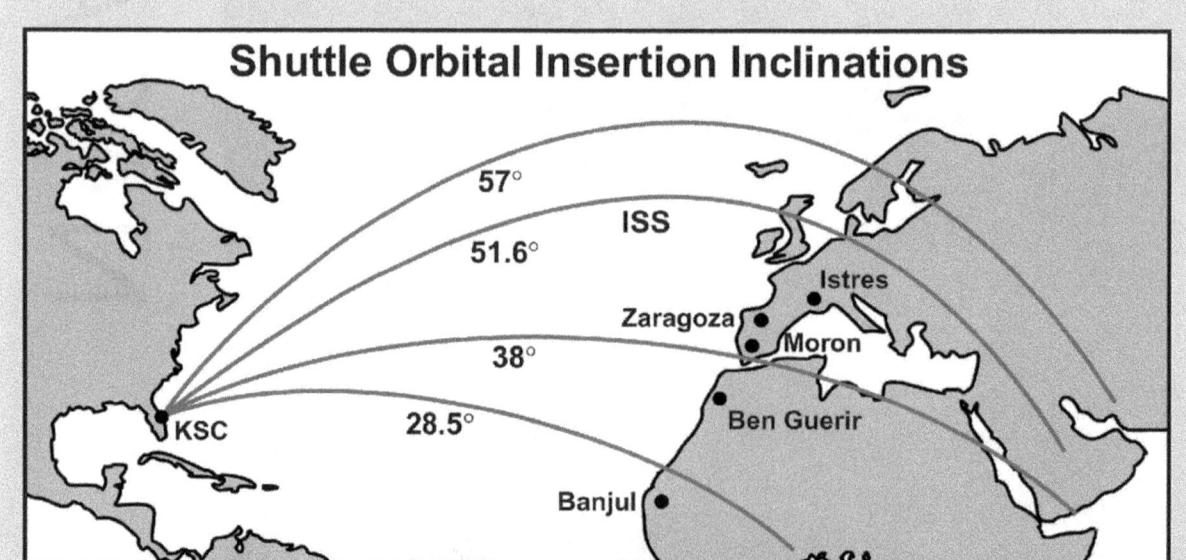

A shuttle liftoff at 57 degrees inclination to the equator is considered high inclination, 38 to 45 degrees is mid-inclination, and 28.5 degrees is low-inclination. TAL sites are selected depending on launch inclination. All flights to the International Space Station are launched on a 51.6-degree inclination.

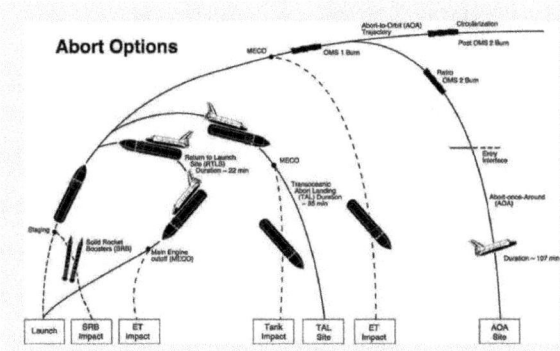

The abort scenario would depend on which point in the liftoff sequence the contingency occurs. Duration refers to the approximate time from liftoff to touchdown.

During a TAL abort, the orbiter continues on a trajectory across the Atlantic to a predetermined runway at one of the TAL sites. The three sites NASA designated as TAL sites were chosen in part because they are near the nominal ascent ground track of the orbiter, which would allow the most efficient use of main engine propellant and cross-range steering capability.

Moron Air Base, Spain

Moron Air Base is a joint-use U.S. and Spanish Air Force Base and was designated a TAL site in 1984. Moron Air Base serves as a weather alternate for low-, mid- and high-inclination launches.

Moron AB is located about 35 miles southeast of Seville and 75 miles northeast of Naval Station Rota. Although Moron is close to the foothills of the Sierra de Ronda mountain chain, most of the surrounding countryside is flat with a few hills and shallow valleys. Elevations vary from 200 to 400 feet above sea level. The weather is generally good with no associated unusual weather phenomena.

The Moron AB runway is 11,800 feet long by 200 feet wide with 50-foot asphalt-stabilized shoulders and 1,000-foot overruns. The runway is equipped with space shuttle-unique visual landing aids, a Microwave Landing System (MLS), a Tactical Air Control and Navigation (TACAN) system and a remote weather tower.

Communications at Moron include three INMARSAT satellite circuits, U.S. Defense Communication Net lines and Spanish commercial telephone lines. Internet capability is available through the base's Large Area Network.

Zaragoza Air Base, Spain

Zaragoza AB was designated a TAL site in 1983 and is the primary TAL site for high-inclination launches. Until the U.S. Air Force pulled out in 1992, it was a joint-use base with a NATO-instrumented bombing range nearby. Today the Zaragoza Spanish Air Force retains the base's status as a TAL site through cooperative agreements between the U.S. government (NASA) and the government of Spain, and between the U.S. Department of Defense (DOD) and the Spanish Ministry of Defense.

Located northwest of the town of Zaragoza, the base has two parallel runways. The civilian airport runway, designated 30R, is 9,923 feet long by 197 feet wide. The Spanish Air Force runway, or space shuttle runway, designated Runway 30L, is 12,109 feet long by 197 feet wide and has 1,000-foot overruns. It is equipped with space shuttle-unique visual landing aids and an MLS, a TACAN system and a remote weather tower.

Through the agreement negotiated between the U.S. and Spanish militaries, NASA has retained the sole use of a hangar complex that is used as the operations and storage building. A building operations and maintenance contractor, with a contract administered out of Moron AB, is permanently stationed at Zaragoza to maintain the NASA/DOD complex and associated ground-support equipment.

Communications at Zaragoza include three INMARSAT satellite circuits and Spanish commercial telephone lines. Internet capability is available through a local Internet service provider.

Istres Air Base, France

Istres AB (known to the French as Base Aerienne 125 or BA-125) was activated during return-to-flight preparations to provide another high inclination site. It is located just outside of the town of Istres in the south of France, approximately 30 miles west of Marseille. The town of Istres, with about 40,000 inhabitants, is situated on the edge of Berre Lake. It is bordered on the south and east by major petrochemical and other heavy industry plants, and on the north and west by small and medium-sized towns, including Aix-en-Provence, Arles, Avignon, Martigues, Salon-de-Provence, Fos-sur-Mer and Marignane.

The Istres area is heavily influenced by its proximity to the major port city of Marseille, the second-largest city in France with a population of around 800,000 inhabitants. Istres AB is a major employer in the local area, with both active French Air Force (FAF) units and French civilian aviation companies located there. The current base population consists of approximately 5,500 military and civilian personnel. The FAF presence includes an air-refueling squadron and a strategic bomber squadron. Istres AB is essentially the equivalent of Edwards AFB, Calif., as the FAF Flight Test Center. The base infrastructure includes those units and activities found on any major air force base.

Istres AB has a single runway oriented in a northwest to southeast direction. It is 197 feet wide with 25-foot shoulders, and is equipped with space shuttle-unique landing aids allowing for landings in the northwest direction only. Runway 33 is the primary runway and is 11,303 feet long with a 1,377-foot underrun and a 3,963-foot load-bearing overrun, for a total usable runway of 15,266 feet. The shuttle-unique landing aids consist of visual landing aids, an MLS, a TACAN system and a remote weather tower. Communications include three INMARSAT satellite circuits and French commercial telephone lines. Internet capability is available through a local Internet service provider.

Banjul, the Gambia

Banjul International Airport in the Republic of the Gambia, West Africa, was the primary TAL site for 28.5-degree, low-inclination launches because of its in-plane location. It was activated in July 1988, replacing a TAL site at Dakar, Senegal, that NASA concluded was unsatisfactory due to runway deficiencies and geographic hazards. Banjul was used for 28 of 36 low-inclination launches before being closed in November 2002.

Ben Guerir, Morocco

The Ben Guerir Air Base in Morocco was used for most of the launches as a weather alternate TAL site because of its geographic location and its landing support facilities. Ben Guerir replaced Casablanca, Morocco, which was used as a contingency landing site in January 1986. Ben Guerir was designated as a TAL site in July 1988 and was last used for STS-111 in June 2002. Ben Guerir is in the process of being closed after supporting 83 missions.

Shuttle Support Equipment at TAL Sites

NASA has enhanced each of the TAL sites with space shuttle-unique landing aids and equipment to support an orbiter landing and turnaround operation. Some of the specific equipment and systems that are installed include the following:

Navigation and Landing Aids

Three navigation aids are used during entry and landing. Beginning at approximately 8 miles from the TAL runway, the Microwave Scanning Beam Landing System or Microwave Landing System will provide highly accurate three-dimensional position information to the orbiter to compute steering commands to maintain the spacecraft on the nominal flight trajectory during the landing phase.

Precision Approach Path Indicator (PAPI) lights are used by the orbiter crew to verify outer glide slope during a landing. Two sets of PAPI lights are used to accommodate high-wind and low-wind scenarios. High-wind PAPI lights are located 6,500 feet prior to the threshold on an extended centerline of the runway, and the low-wind PAPI lights are located 7,500 feet prior to the threshold on the centerline.

Ball/bar lights are used by the space shuttle astronauts to verify the proper inner glide slope during landing. The ball/bar lights are installed along the runway on the left, which is the commander's side of the orbiter. The ball light is located 1,700 feet down the runway from the threshold with the bar light at 2,200 feet. Superimposing the ball light on the bar lights places the orbiter on a 1.5-degree glide slope and enables the orbiter crew to touch down approximately 2,500 feet down the runway.

Distance-to-go markers display to the crew the distance remaining to the end of the runway during landing and rollout. These markers are installed on the left side of the runway, 1,000 feet apart, starting from the threshold and counting down to the overrun.

Xenon lights are high-intensity flood lights that provide runway lighting for night landings at the TAL sites. Each light provides 1 billion candle power each. A set of three lights is installed on raised platform trucks on each side of the runway at the beginning of the underrun, shining down the runway to provide illumination of the entire touchdown area.

Portable approach lights (flashlights) are required for night landings at Istres and Zaragoza because no approach light systems are installed on the space shuttle runway approach paths. The flashlights are placed in a predetermined pattern on the underrun and along a 3,000-foot extension of the runway centerline to give a lighted visual reference of the approach path to the runway.

The PAPI lights, Xenon lights and portable approach lights are installed prior to each space shuttle launch, dismantled after the TAL site is released from support, and stored until required for the next mission.

Weather Equipment

Each TAL site has an automated weather station or tower that collects and transmits weather data every four hours, 365 days a year, via satellite to the Spaceflight Meteorology Group (SMG) at Johnson Space Center in Houston. Responsibility for weather forecasting for the space shuttle program rests with the SMG.

The DOD deploys U.S. Air Force or Navy weather personnel to the TAL sites to provide real-time weather observations from launch minus 48 hours to launch plus 30 minutes. These personnel act as the SMG weather point of contact on site and provide hourly weather observations to the SMG to assist in accurately forecasting weather conditions at the TAL sites. They also operate the TAL Atmospheric Sounding System (TASS), ceilometers and visibility detectors installed at the sites. Ceilometers measure the cloud ceiling while a visibility detector provides information on the amount of dust in the air.

The TASS automatically tracks weather instruments called "rawindsondes" that are carried aloft by weather balloons to monitor upper winds and other data. This data is transmitted to the SMG via the TAL INMARSAT satellite circuits and/or commercial telephone lines. Flight rules call for at least one TAL site to be in the "go" status for weather, meaning it would be suitable for an orbiter landing, before a space shuttle launch will be made from Kennedy Space Center.

Dedicated Ground-Support Equipment

Dedicated orbiter ground-support equipment has been prepositioned at the TAL sites. This equipment includes a hatch opening tool, tow bar, tow bar adapter, staircase for the crew to disembark from the orbiter, grounding cable, landing gear lock pins, tire chocks, light banks for night operations and many

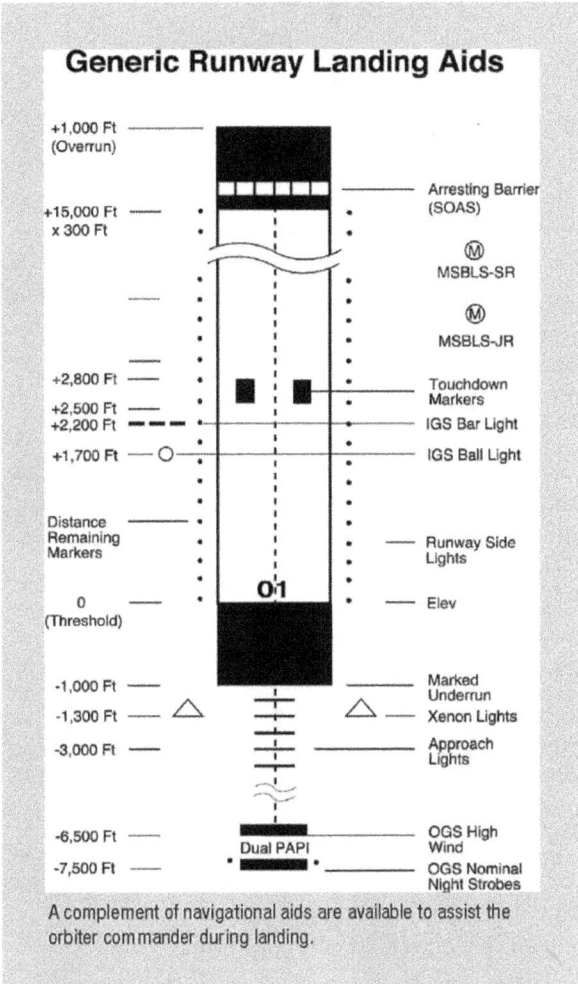

A complement of navigational aids are available to assist the orbiter commander during landing.

more pieces of equipment for ground support.

Extra tires, brake-removal equipment and a Rhino jack – used for jacking up the orbiter – are prestaged at Moron AB. Since a C-130 is not staged at Moron, this equipment would be moved to the actual TAL landing site by a C-130 aircraft coming from Zaragoza AB, Spain, or Istres, France.

Emergency Equipment

Fire, crash and rescue (F/C/R) resources include firefighting equipment and personnel. A team of seven Air Force F/C/R personnel from Europe deploys to the Spanish TAL sites for contingency landing support and is augmented by 18 trained firefighters from the host country. The French Air Force provides both internal rescue and 18 external firefighters for support at Istres.

Aircraft Support

Aircraft support at the TAL sites, and all other DOD support to the space shuttle program, is managed through the DOD's Manned Space Flight support office, located at Patrick Air Force Base, Fla. A C-130 aircraft is deployed to the Zaragoza and/or Istres TAL site two days prior to launch.

The C-130 serves a variety of roles, including search and rescue, medical evacuation and logistics. The TAL site C-130s are equipped with eight crew members, three air-deployable Zodiac rafts, nine pararescue jumpers, two DOD flight surgeons, a nurse and medical technician, and approximately 2,500 pounds of medical equipment.

The TAL sites are also supported by a DOD weather aircraft, either a C-21 (similar to a Learjet) or a C-12 (Beachcraft turboprop). An astronaut flies on this aircraft to provide real-time weather observations for the Spaceflight Meteorology Group and recommend go/no go status to the flight director at Johnson Space Center.

The astronaut is referred to as the TALCOM, the TAL site equivalent of the CAPCOM, or capsule communicator. The CAPCOM is the Mission Control-based astronaut in Houston who serves as the communication liaison with on-orbit space shuttle crews.

TALCOMs are deployed to each of the three TAL sites supporting a launch as the Johnson Flight Crew Operations Directorate representative; at the TAL site, he or she is also designated as the deputy ground operations manager.

The TALCOM is normally airborne from T-1:30 hours (one hour, 30 minutes before launch) through main engine cutoff. The aircraft's UHF radio is linked to the weather CAPCOM and the Spaceflight Meteorology Group at Mission Control.

The TALCOM also becomes familiar with the surrounding terrain along the approach path to the runway at the TAL site, and his or her observations are duly noted to assist an orbiter commander during a landing. The TALCOM checks out slant-range visibility and intensity settings on the visual landing aids, PAPI and ball/bar lights.

Preparing for a TAL

Seven or eight days prior to a space shuttle launch date, depending on the TAL site, a team of NASA and contractor personnel will depart Kennedy Space Center and begin activating the TAL sites assigned to support the mission. Four to five days are required on site to prepare the TAL site for launch support.

The mission support team is managed by the NASA ground operations manager and includes about 20 contractor personnel. DOD support for NASA and the manager include a deployed forces coordinator and two aircraft with an additional 35 personnel. The majority of the DOD personnel arrive on site 48 hours prior to scheduled liftoff.

If a TAL were declared, the ground operations manager at the TAL site would be notified by the landing support officer in the Mission Control Center at Johnson Space Center that the space shuttle was aborting to the emergency landing site. The landing support officer would begin coordination to clear the upper air space with the Federal Aviation Administration and the International Civil Aviation Organization. The U.S. State Department would notify the American embassy in the country involved.

The time from declaration of a TAL abort to a landing is estimated at approximately 25-30 minutes. Once the space shuttle crew commander selects the TAL option, the preprogrammed onboard orbiter computers would automatically steer the craft toward the designated landing site. The orbiter would roll heads up before main engine cutoff and all extra fuel would be dumped to increase vehicle performance by decreasing weight and reducing the toxic environment in and around the orbiter after a landing.

The space shuttle would be flown to an altitude of about 350,000 feet and the main engines would cut off at the correct velocity. The external tank would be jettisoned after main engine cutoff, as in a normal launch, and tumbled to ensure that it burns up on re-entry into the Earth's atmosphere.

A preloaded re-entry program would then go into effect, with the orbiter encountering the atmosphere and a normal re-entry planned. Ten minutes before landing, communications would resume through the Tracking and Data Relay Satellite network, used for orbiter/Mission Control contact.

The landing support officer and flight director in the Mission Control center would keep the ground operations manager and the TAL site informed of the status of the orbiter systems during the approach to the TAL site. Data received from the TAL site TACAN would be used to update the orbiter's inertial guidance system 200 miles from touchdown as the spacecraft slowed to Mach 7 (seven times the speed of sound).

At landing minus six minutes, the orbiter would enter what is referred to as the terminal area. At this point, its altitude is still quite high (82,000 feet) and its speed still supersonic at Mach 2.5. Its flight would be akin to a conventional aircraft's except that the orbiter's speed brakes would be left open to provide greater stability during supersonic flight.

Approximately five minutes before touchdown, the orbiter's speed would be approximately Mach 1. About four minutes before touchdown, the commander would take over manual control of the spacecraft. This would be just prior to a maneuver known as intercepting the Heading Alignment Circle.

The Heading Alignment Circle is a large turn to align the orbiter with the centerline of the runway and to allow the commander to bleed off or reduce any excess speed the vehicle may have. At landing minus two minutes, the orbiter would enter its final approach at an altitude of 13,000 feet. The speed brakes would be closed at an altitude of 3,000 feet.

At an altitude of 1,800 feet and a distance of 7,500 feet from the threshold of the runway, the commander would begin a preflare maneuver to pull up from a glide slope of 19 degrees to a gentler one of 1.5 degrees. Touchdown normally would occur at a speed of about 200 miles per hour.

A typical power-down would be completed before the crew members exited the orbiter, much the same as in a normal end-of-mission landing. At a TAL site, this would take approximately 30 minutes to accomplish.

At about T+ 3 hours (touchdown plus three hours), the crew members would depart the TAL site onboard the C-130 aircraft en route to the hospital at Naval Station Rota, Spain (if uninjured or with minor injuries), where they would be met by the crew-return aircraft from Johnson Space Center for their return to the United States.

If there were severely injured crew members, they would be medevaced on the C-130 aircraft or taken by ambulance to identified critical care medical facilities in Europe. The crew would remain together, escorted by security forces, unless medical circumstances or aircraft availability dictated otherwise.

Post-Landing Operations

Once the crew exited the orbiter, and the recovery management team at Kennedy Space Center granted permission, the crew hatch would be closed and the orbiter prepared for towing to a remote deservicing area or park site. Safing and deservicing of the orbiter would be initiated by the deployed TAL team and augmented by a team known as the Rapid Response Team.

A Mishap Investigation Team may also travel to the TAL site to collect data and conduct a mishap investigation on the unscheduled landing. The TAL site ground operations manager would initially be in charge until relieved by a higher-ranking management official who would arrive on the response/investigation team aircraft.

Within 24 hours, the response/investigation teams would arrive at the TAL site aboard C-17 aircraft carrying personnel and equipment. Most of the equipment would come from Kennedy Space Center and Dryden Flight Research Center in California.

Following the advance response/investigation teams contingent, the Deployed Operations Team, consisting of additional personnel and equipment, would begin arriving at the TAL site for the orbiter turnaround operation. NASA estimates it would take about 19 C-17/C-5 aircraft sorties, a significant Navy sealift operation, and 450 NASA and contractor personnel to complete the turnaround.

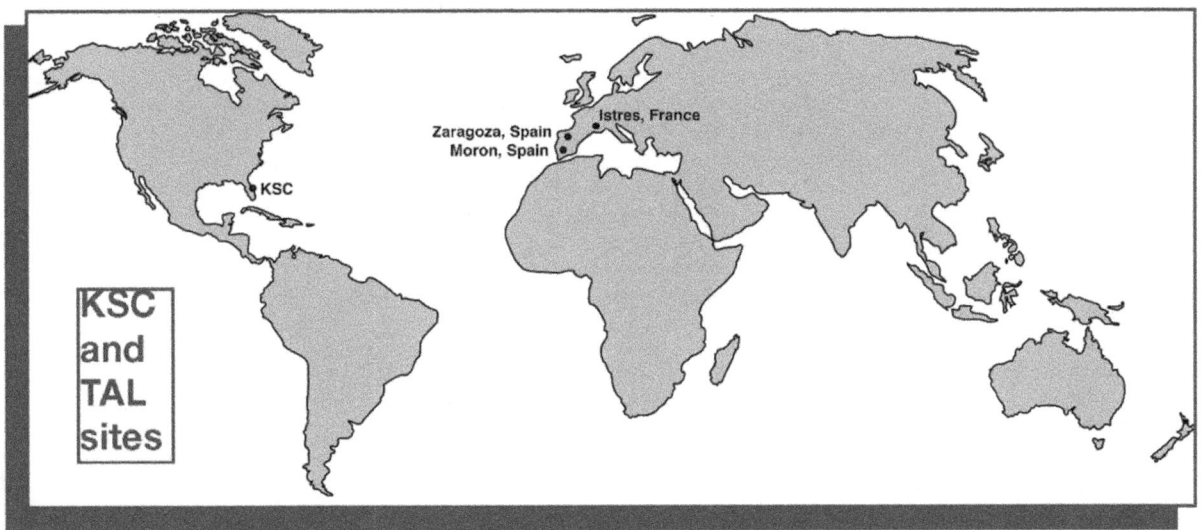

Not all of these personnel would be on site at any one time.

In addition to these personnel, another 150 to 200 DOD personnel may be required to put in place a "bare-base" operation consisting of portable general-purpose shelters, latrines, a kitchen, aircraft hangars and other support equipment if the TAL site does not have adequate facilities to support such a large team.

Payloads and/or airborne support equipment will remain onboard the orbiter for the flight back to Kennedy Space Center unless the capability of the shuttle carrier aircraft, landing site location or other requirements dictate otherwise.

Kennedy Space Center, home of the space shuttle, is making the Vision for Space Exploration a reality.

National Aeronautics and Space Administration

John F. Kennedy Space Center
Kennedy Space Center, Fla.

www.nasa.gov

PROJECT MERCURY

FAMILIARIZATION MANUAL
Manned Satellite Capsule

Periscope Film LLC

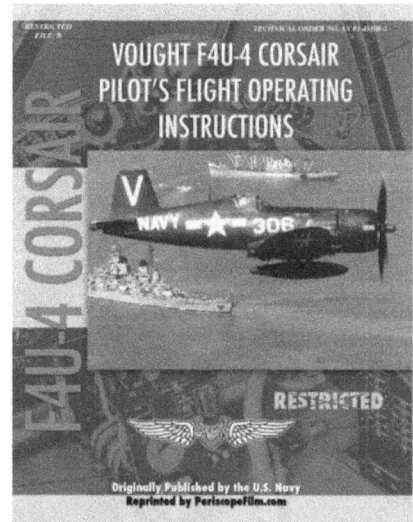
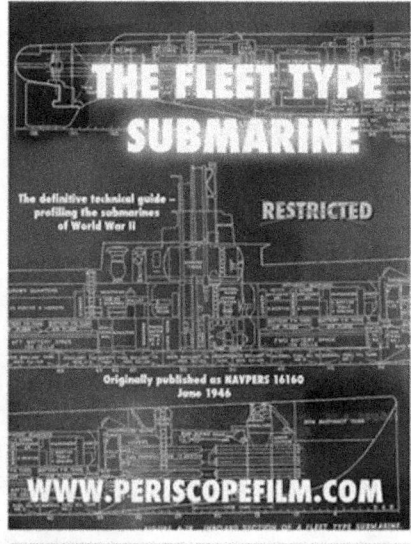
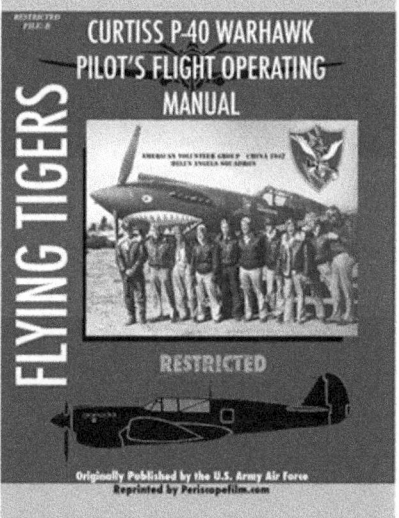

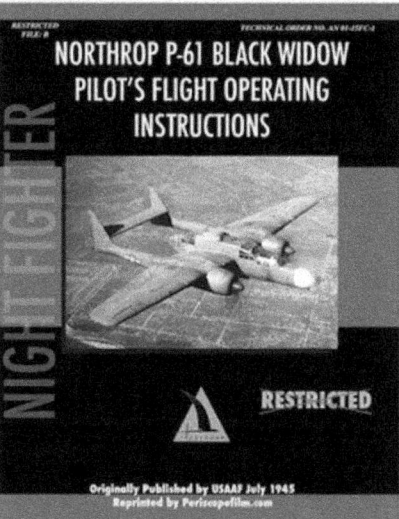

ALSO NOW AVAILABLE
FROM PERISCOPEFILM.COM

©2012 Periscope Film LLC
All Rights Reserved
ISBN#978-1-937684-78-5
www.PeriscopeFilm.com

www.ingramcontent.com/pod-product-compliance
Lightning Source LLC
Chambersburg PA
CBHW081839170426
43199CB00017B/2778